GENES in the FIELD

On-Farm Conservation of Crop Diversity

GENES in the FIELD

On-Farm Conservation of Crop Diversity

Edited by

Stephen B. Brush, Ph.D.

INTERNATIONAL PLANT GENETIC RESOURCES INSTITUTE
ROME, ITALY

IDRC
INTERNATIONAL DEVELOPMENT RESEARCH CENTRE
CANADA Ottawa, Canada

LEWIS PUBLISHERS

Boca Raton London New York Washington, D.C.

Library of Congress Cataloging-in-Publication Data

Genes in the field: on-farm conservation of crop diversity / edited by Stephen B. Brush.
      p. cm.
     Includes bibliographical references and index.
     ISBN 0-88936-884-8 International Development Research Centre
     ISBN 1-56670-405-7 Lewis Publishers (alk. paper)
1. Crops--Germplasm resources. 2. Germplasm resources, Plant. I. Brush, Stephen B. Brush. 1943–
SB123.3.G47 1999   99-044933
  631.5'23--dc21   CIP

Canadian Cataloguing in Publication Data

Main entry under title:

Genes in the field : on-farm conservation of crop diversity

Copublished by International Plant Genetic Resources Institute.
Includes bibliographical references and index.
ISBN 0-88936-884-8

1. Crops – Germplasm resources.
2. Crops – Genetic engineering.
3. Germplasm resources, Plant.
4. Plant diversity conservation.
I. Brush, Stephen B., 1943–
II. International Plant Genetic Resources Institute.
III. International Development Research Centre (Canada).

SB123.3G46 1999   631.5'23'3   C99-980391-3

Copublished by
Lewis Publishers
2000 N. W. Corporate Blvd.
Boca Raton, FL 33431
U.S.A.

International Development Research Centre
P. O. Box 8500
Ottawa, ON
Canada K1G 3H9

and by
International Plant Genetic Resources Institute
Via delle Sette Chiese 142
00145 Rome, Italy

© 2000 International Development Research Centre and International Plant Genetic Resources Institute
Lewis Publishers is an imprint of CRC Press LLC

No claim to original U.S. Government works
International Standard Book Number 0-88936-884-8
International Standard Book Number 1-56670-405-7
Library of Congress Card Number 99-044933
Printed in the United States of America 1 2 3 4 5 6 7 8 9 0
Printed on acid-free paper

Contents

Section IV.   Policy and institutional issues

Foreword

In the 1920s and 1930s, pioneering scientists such as N.I. Vavilov and Harry Harlan began to notice that traditional crop varieties, or landraces, were being lost from fields and gardens around the world. For the next 60 years, scientific efforts to conserve plant genetic diversity focused on collecting material and placing it in *ex situ* storage. Institutions were created, gene banks were constructed, and millions of accessions were accumulated and deposited in cold stores.

It was not until the 1980s and 1990s that we took full account of the fact that a great deal of diversity still existed, uncollected and unfrozen, but "conserved" nevertheless, under the care and management of farmers and gardeners. A professional community of scientists, plant breeders, and others had created the capacity for conserving a vast amount of diversity in *ex situ* conditions. But farmers — particularly women in developing countries — had persisted in managing and developing their planting materials. *In situ* conservation was a necessary step and by-product of the millennia-old practice of sowing and reaping and preparing to do so again the next season.

Appreciating the significance and value of something that has been present for so long is never an entirely smooth or comfortable process. Terminology — and the encrusted notions it sometimes hides — is challenged straightaway. Stone Age varieties. Primitive varieties. Landraces. Heirloom varieties. Traditional cultivars. Farmer varieties. All of these terms have been used to describe the same materials. Similarly, different terms have been employed to describe the process itself. Is "it" *in situ* conservation, dynamic conservation, or on-farm management and improvement, for example?

Whatever "it" is, this activity is now firmly ensconced in our consciousness, not to mention institutional programs. The Convention on Biological Diversity and the FAO Global Plan of Action for Plant Genetic Resources have given both legitimacy and prominence to *in situ* or on-farm efforts. The present volume will not solve the problem of terminology or resolve the different perspectives of different authors: the reader must accept that different authors see and describe the world differently. Instead, contributions have been solicited with the aim of solidifying and extending our knowledge of what is taking place — and what could take place — in the field.

Approximately 1.4 billion people live in farm families that are largely self-reliant and self-provisioning for their seeds and other planting materials.

Farmers may see advantages in maintaining these materials. Or there may be no alternatives that can meet their needs, however defined. What is clear is that diversity — despite continuing threats — will continue to exist in farms and fields around the globe for the foreseeable future.

If it is now accepted that continued production and use of local cultivars on farms can play a significant role in the conservation of within-species genetic diversity, it must also be acknowledged that methodologies and practices are neither fully understood nor elaborated. Indeed, the concept of on-farm conservation raises an enormous number of largely unresolved questions. There are questions concerning the effectiveness of on-farm management by farmers and local communities as a reliable mechanism for the conservation of specific local cultivars for the larger community of plant breeders and users. There are further questions concerning the role of farmers and the ways in which their needs and interests can be recognized and benefits from their work realized. There are also questions concerning the interplay between conservation and production and the progress of agricultural development. And there are questions about the proper role and relationship of farmers and scientists in crop improvement and diversity conservation endeavors. All of these topics are undergoing scrutiny and lively debate.

The views expressed in this volume reflect differences in the authors' backgrounds, experiences, and interests. Those concerned primarily with farmers' needs and interests for maintaining certain production levels and meeting food and income needs may have quite different concerns from those whose focus is on the maintenance of viable and sustainable ecosystems, or those whose interest lies with the maintenance of maximum levels of potentially useful genetic diversity. Exploring and understanding these different concerns is an essential starting point for answering some of the key questions about the implementation of on-farm conservation and the role of local cultivars in sustainable development.

This book provides an opportunity for various authors from widely differing backgrounds to explore some of the issues raised by conserving and improving crops *in situ*. A wide range of expertise and experiences will be needed to develop realistic approaches to conservation on farms. Genetic, ecological, agricultural, social, economic, and legal concerns all have to be considered and integrated in developing practical work plans at international, national, and local levels. In this book, authors with experience in different fields explore some of the problems and possibilities from their perspective.

It is no longer necessary to ask whether *in situ* conservation of crop plants should be undertaken but rather to discuss how, when, and where it is done, and how it might be enhanced. Definite answers may be scarce. Our ambition, however, is to take the discussion a step forward and to provide a framework for discussing the many problems still to be resolved.

Cary Fowler
Geoffrey C. Hawtin
Toby Hodgkin

Contributors

Zemede Asfaw, Ph.D.
Professor
Department of Biology
Addis Ababa University
P.O. Box 1176
Addis Ababa
Ethiopia

Anthony H.D. Brown, Ph.D.
Senior Principal Research
  Scientist
CSIRO Plant Industry
GPO Box 1600
Canberra, ACT 2601
Australia

Stephen B. Brush, Ph.D.
Professor
Department of Human and
  Community Development
University of California
Davis, CA 95616

Salvatore Ceccarelli, Ph.D.
Barley Breeder
International Center for
  Agricultural Research in
  Dry Areas (ICARDA)
P.O. Box 5466
Aleppo
Syria

Carlos M. Correa, Ph.D.
Master Programme on Science and
  Technology and of the
  Postgraduate Course on
  Intellectual Property
University of Buenos Aires
Monasterio 1138
Vte. Lopez 1638
Argentina

Elizabeth Cromwell, M.Sc.
Research Fellow
Overseas Development Institute
Portland House
Stag Place
London SW1E 5DP
United Kingdom

Regassa Feyissa, M.Sc.
Institute of Biodiversity
  Conservation and Research
Addis Ababa
Ethiopia

Cary Fowler, Ph.D.
Honorary Research Fellow
International Plant Genetic
  Resources Institute (IPGRI)
Via delle Sette Chiese 142
00145 Rome
Italy

Timo Goeschl, D.Phil.
Marie Curie Research Fellow
Faculty of Economics and Politics
Cambridge University
Cambridge, CB3 9DD
United Kingdom

Tirso A. Gonzales, Ph.D.
University of California President's
   Postdoctoral Fellow
Department of Native American
   Studies
University of California
Davis, CA 95616

Stefania Grando, Ph.D.
Barley Breeder
International Center for
   Agricultural Research in Dry Areas
   (ICARDA)
P.O. Box 5466
Aleppo
Syria

Geoffrey C. Hawtin, Ph.D.
Director General
International Plant Genetic
   Resources Institute (IPGRI)
Via delle Sette Chiese 142
00145 Rome
Italy

Toby Hodgkin, Ph.D.
Principal Scientist
International Plant Genetic
   Resources Institute (IPGRI)
Via delle Sette Chiese 142
00145 Rome
Italy

Devra Jarvis, Ph.D.
Scientist
International Plant Genetic
   Resources Institute (IPGRI)
Via delle Sette Chiese 142
00145 Rome
Italy

Dominique Louette, Ph.D.
Agronomist
Instituto Manantlán de Ecología y
   Conservación de la Biodiversidad
   (IMECBIO)
Universidad de Guadalajara
   (UdeG)
A.P. 64
151 Av. Indpendencia Nacional
48900 Autlán, Jalisco
México

Timothy Swanson, Ph.D.
Professor of Economics
School of Public Policy
University College London
London, WC1E 6BT
United Kingdom

Tesfaye Tesemma, Ph.D.
Debre Zeit Agricultural Research
   Center
Alemaya University of Agriculture
Debre Zeit
Ethiopia

Saskia van Oosterhout, Ph.D.
Sorghum Landrace Study
Agriculture Research Centre
P.O. Box CY 594
Causeway
Harare
Zimbabwe

Melaku Worede, Ph.D.
International Scientific Advisor
Seeds of Survival
P.O. Box 2525
Addis Ababa
Ethiopia

Acknowledgments

A primary goal of agricultural research is to develop technology that will enable the world's farmers to produce enough food in a manner that is sustainable and economically viable. Agricultural scientists have long acknowledged their debt to farmers in the joint enterprise of developing new technology. This debt extends not only to our contemporaries but also to past generations. The concept of *in situ* conservation is partly an effort by the scientific community to honor its debt to the legacy of farming peoples who created the biological basis of crop production. So, the first acknowledgment of this book is to the farmers who shared their time, resources, and knowledge with the authors of these chapters. It is hoped that their generosity and skills are fairly reflected.

The support of the International Development Research Centre of Canada has been crucial in bringing this book to fruition. It is hoped that this book appropriately reflects the IDRC's foresight and innovation in matters pertaining to agricultural development. Likewise, the staff members of the International Plant Genetic Resource Institute have been leaders in supporting different conservation methods for the world's storehouse of crop genetic resources. In particular, Toby Hodgkin has shown imagination and effectiveness in moving *in situ* conservation into the mainstream of agricultural conservation. The genesis of this book is a product of Toby's integrity and determination.

Karyn Fox of the International Agricultural Development program at the University of California, Davis was most directly responsible for the production of this volume and the editing of its contents. Our debt to her patience, persistence, intelligence, and good humor is immeasurable. The future of participatory research and development programs appears to be secure in the hands of Karyn's generation.

Section I

Introduction and overview

chapter one

The issues of in situ conservation of crop genetic resources

Stephen B. Brush

Introduction

Domesticated plants have been fundamentally altered from their wild relatives; these species have been moved into and adapted to new environments; they have become dependent on the tiller's hand; and they have been reshaped to meet human needs and wants. Modern crops are the result of thousands of years of these evolutionary processes. Like all biological evolution, crop evolution involves two fundamental processes: the creation of diversity and selection (Harris and Hillman 1989). Crop evolution is distinguished by two types of selection: one natural and another artificial or conscious. These evolutionary processes must continue in order for agriculture, a living and evolving system, to remain viable. Therefore, an essential criterion of crop evolution is the availability of genetic diversity. Crop evolution has been altered by our enhanced ability to produce, locate, and access genetic material, but this has not changed its fundamental nature. Both farmers and scientists have relied on the store of genetic diversity present in crop plants that has been accumulated by hundreds of generations who have observed, selected, multiplied, traded, and kept variants of crop plants. The result is a legacy of genetic resources that, today, feeds billions of humans.

Genetic diversity is important both to individual farmers and farming communities and to agriculture in general. Individual farmers value diversity within and between their crops because of heterogeneous soils and production conditions, risk factors, market demand, consumption, and uses of different products from an individual crop species (Bellon 1996). Thus a wheat farmer in Turkey may have different types of wheat for hillside or valley bottom areas, for irrigated and rain-fed parcels, for homemade bread and for urban grain markets, for straw and animal feed (Brush and Meng 1998). Moreover, farmers usually rely on diversity of other farms and communities to provide new seed when crops fail or seed is lost or to renew seed that no longer meets the farmer's criteria for good seed (Louette et al. 1997). The need for diversity at both the farm and regional levels has resulted in a vast store of genetic diversity in crops, a store passed down from earlier generations and maintained for the future. In regions where a crop's evolution has the longest record, where the crop was originally domesticated, and where its diversity is greatest, the local store of genetic diversity in farming communities is also a store of genetic resources for that crop, an invaluable resource for farmers, scientists, and consumers elsewhere (Hawkes 1983).

Unfortunately, this legacy is imperiled by the very conditions it helped to create (Wilkes 1995). Record numbers of humans, agricultural science and technology, and economic integration of the world's many diverse cultures threaten to destroy this legacy, as modern crop varieties and commercial farming diffuse into every agricultural system. A result of these changes is that diversity on individual farms and across wide regions is threatened by modern crop varieties that have been bred for broad adaptation, resistance to disease and other risk factors, ability to better use water and fertilizer, and higher yields. This threat is evidenced by the fact that agricultural development in Europe, North America, and many less developed countries has been accompanied by the replacement of diverse, local populations of crops with a handful of modern varieties.

The importance of crop genetic resources and threats to them has led to the creation of conservation programs to preserve crop resources for future generations. One type of crop genetic conservation is *ex situ* — maintenance of genetic resources in gene banks, botanical gardens, and agricultural research stations (Plucknett et al. 1987). Another type is *in situ* — maintenance of genetic resources on-farm or in natural habitats (Brush 1991; Maxtel et al. 1997a). In actuality, two types of *in situ* conservation can be distinguished. First, *in situ* conservation refers to the persistence of genetic resources in their natural habitats, including areas where everyday practices of farmers maintain genetic diversity on their farms. This type is a historic phenomenon, but it is now especially visible in regions where farmers maintain local, diverse crop varieties (landraces), even though modern, broadly adapted, or higher yielding varieties are available.

Second, *in situ* conservation refers to specific projects and programs to support and promote the maintenance of crop diversity, sponsored by national governments, international programs, and private organizations. *In situ* conservation programs may draw on the existence and experience of the first type, but they are designed to influence farmers in the direction of maintaining local crops by employing techniques that may not be local. This type of conservation faces daunting tasks. It must cope with continual social, technological, and biological change while preserving the critical elements of crop evolution — genetic diversity, farmer knowledge and selection, and exchange of crop varieties.

In situ conservation practices and projects in agriculture theoretically can concern the wide spectrum of genetic resources relating to crops, from wild and weedy relatives of crop species to the infraspecific diversity within crop species (Maxted et al. 1997b). The focus of this chapter and the others in this book is the latter, the diversity within cultivated species, exemplified by heterogeneous crop populations known as landraces. These are named, farmer varieties that usually have a reduced geographic range, are often diverse within particular types, and are adapted to local conditions (Brush 1995; Harlan 1995). One reason for our focus on diversity within cultivated crops is that science of *in situ* conservation of cultivated resources is relatively less developed than the science of conserving wild resources such as wild and weedy crop relatives. Another reason is that *in situ* conservation of cultivated plants requires novel approaches, while *in situ* conservation of wild crop relatives can draw on theories and methods developed for conserving many different species in their natural habitats. Finally, focusing on variation within cultivated species is warranted by the fact that this type of diversity is arguably the most important one for the future viability of agricultural evolution, as it has been in the past.

The successful planning and implementation of projects for on-farm (*in situ*) conservation of crop genetic resources require us to answer four questions. First, why undertake this type of conservation, especially when investments are made for *ex situ* conservation? Second, what scope is necessary or appropriate for *in situ* conservation of crop germplasm? Third, how can agricultural agencies and organizations promote this form of conservation? Finally, what legal and institutional questions pertain to on-farm conservation of genetic resources? The answers to these questions come from different fields of science, for example, population biology and social science, and from law and politics. Moreover, the answers to these questions seldom are definitive. More important than definitive answers is the ability to seek answers, because new answers will be needed for different times, conditions, crops, and societies. The purpose of this and other chapters in this book is not to answer these four questions but rather to offer guideposts and a context for finding answers in specific regions and for specific crops and cropping systems.

Why in situ *conservation?*

The invention and development of agriculture was accomplished independently in several places in the world, but within a relatively narrow time period following the end of the Pleistocene period — 8,000 to 10,000 years before the present (Harris and Hillman 1989). Why agriculture arose during this limited time period and only in a few places, and exactly how wild plants were identified, manipulated, and managed for domestication remain mysteries. Although the origins and processes of crop domestication are obscure, its consequences are well known and thoroughly documented —

the creation of an entirely new way of life and eventual rise of urban civilization with all of its wonders and woes. Since the time of domestication, a progression of changes has occurred in farming systems and social systems associated with agriculture. Greater numbers of people than ever before in human history are dependent on a smaller number of crop species; a handful of "mega-crops" have supplanted locally important crops and now feed most of the world's population (Wilkes 1995). The reduction in interspecies diversity of food plants continues the trend of exercising ever greater control over nature and the production process, a trend also supported by the increased use of manufactured inputs in crop production.

Individual social and production systems have been gradually but inexorably integrated into a single, interconnected world system of economic, cultural, and technology exchange, and this integration threatens genetic diversity of crops as much as population increase and modern technology. Until recently, most crop production was intended for local consumption, and it relied mostly on local resources of energy and crop germplasm. Today, however, exceedingly few farming systems function in isolation from markets, national and international political influence, and flows of capital, energy, and technology. Although most farmers still produce their own food, they also sell an appreciable amount into local and national markets. The use of non-local technology and inputs, such as fertilizers, pesticides, and mechanization, is ubiquitous. An increasingly important part of the flow of technological goods to farmers is improved crop varieties, selected from outstanding farmer varieties, developed and released by public crop improvement programs, or sold by private seed companies.

The economic, political, and technological integration of farming systems is generally seen as a positive step that enables development — increased production, income, and well-being (Hayami and Ruttan 1985). Nevertheless, this integration has several negative impacts. Farmers relinquish personal and local control of the production system as they become subject to market and political systems that are not always stable or positive for particular locations or commodities (Chambers 1983; Cernea 1985). Communities and farming systems may become more stratified economically. Increasingly uniform crops may be more vulnerable to pests and diseases. Local knowledge and crop diversity may be lost because of the diffusion of improved, exotic technology. These negative impacts may be ameliorated by policy and technological means, although the knowledge and ability to manage the negative impacts of change are often underdeveloped. Nevertheless, it is important to note that lack of socioeconomic integration also carries potentially serious negative impacts, especially given population growth.

Cultivar diversity in association with wild or ancestral crop species is linked to crop domestication and, most importantly, a broad base of genetic resources that may be useful for crop improvement. The loss of crop varieties from centers of diversity causes genetic erosion or a loss of genetic resources — a negative consequence of agricultural development. Natural historians and biologists have long recognized that particular areas harbored unusually diverse and rich stores of crop germplasm (Harris 1989). One contribution of N. I. Vavilov (1926) was to perceive that these stores were important resources for crop improvement. Shortly after Vavilov's observation, it was noted that these concentrations of crop germplasm were vulnerable to loss, as technological and economic change occur (Harlan and Martini 1936). Once the stores of crop germplasm were identified, a worldwide effort was initiated, first to sample and then to conserve the genetic diversity of major food staples (e.g., rice, wheat, maize, potato, cassava, sorghum, millet, barley, common bean, soybean). The conservation effort focused on preserving crop germplasm that is held in the thousands of distinct crop varieties or cultivars. By 1980, a large portion of the estimated diversity of major staples had been collected for preservation in *ex situ* facilities — gene banks, botanical gardens, and working collections of crop scientists.

During the establishment of the current gene conservation effort (1970–1980), *in situ* conservation was perceived as a possible alternative strategy for conserving crop germplasm, yet it was dismissed for several reasons (Frankel 1970). Most importantly, it was assumed that progress in achieving economic development in diverse agricultural systems inevitably requires the replacement of local crop populations with improved ones. Because genetic diversity in crops is associated with traditional agricultural practices, it is also linked to underdevelopment, low production, and poverty. The positive relationship between crop diversity and poverty is seemingly confirmed by the fact that agricultural development in many places and at different times occurred with the replacement of local and diverse crops, for example, in the hybrid maize revolution in U.S. agriculture between 1920 and 1950 (Cochrane 1993). A corollary of the relationship between diversity and poverty is that conserving traditional crops and their genetic diversity on-farm is tantamount to trying to stop agricultural development. Another reason for rejecting *in situ* conservation is the assumption that farmers who grow traditional crop varieties would require a direct monetary subsidy to continue this practice once improved varieties become available. Such subsidies are not only expensive but also unreliable and difficult to manage for any length of time. Finally, crop scientists who promoted conservation were not interested in conservation alone but also in using genetic resources for crop improvement. As long as breeders' work is confined to experiment stations and laboratories, genetic resources that remain in farmers' fields are not directly useful for crop improvement.

Several decades of collection and gene bank storage of crop genetic resources and research on agricultural change under modern conditions have changed the views that led to the dismissal of *in situ* conservation in favor of *ex situ* methods (Maxted et al. 1997a). One important shift in attitudes is the view that *in situ* and *ex situ* methods are no longer perceived as exclusive alternatives to each other. Today, they are seen as complementary approaches rather than as rivals. There is recognition that these methods address different aspects of genetic resources, and neither alone is sufficient to conserve the total range of genetic resources that exist. Second, it is evident that traditional agriculture and genetic diversity are not inexorably linked and that agricultural development is not incompatible with the on-farm maintenance of diversity. Third, a variety of methods, apart from direct financial subsidies, are available to promote the maintenance of crop genetic resources by farmers.

Five reasons can be cited for promoting *in situ* conservation of crop genetic resources:

1. Key elements of crop genetic resources cannot be captured and stored off-site.
2. Agroecosystems continue to generate new genetic resources.
3. A backup to gene bank collection is necessary.
4. Agroecosystems in centers of crop diversity/evolution provide natural laboratories for agricultural research.
5. The Convention on Biological Diversity mandates *in situ* conservation.

Key elements maintained by in situ *conservation*

The complementarity of *in situ* and *ex situ* conservation is based on the recognition that crop genetic resources involve more than the alleles and genotypes of crop populations. Besides the genetic raw material of landraces, crop genetic resources also comprise related species, agroecological inter-relationships, and human factors. Wild and weedy relatives of crops, as well as perennials and species with recalcitrant seeds, have been recognized as elements of crop genetic resources that cannot be contained in *ex situ* facilities. In addition, we now recognize that ecological relationships such as gene flow between different populations and species, adaptation and selection to predation and disease, and human selection and management of diverse crop resources are components of a common crop evolutionary system that generate crop genetic resources. The broader ecological view of crop genetic resources, then, includes not only alleles and genotypes of diverse crop populations but also wild and weedy crop relatives, predators and diseases, and systems of agricultural knowledge and practice associated with genetic diversity (Altieri and Merrick 1987).

While *ex situ* conservation is well suited to capture and store alleles and genotypes, it is not suited to the conservation of the other components of the agroecosystem that generate crop genetic resources. *In situ* conservation is specifically intended to maintain those components in living, viable agroecosystems. A critical difference between *ex situ* and *in situ* conservation is that the former is designed to maintain the genetic material in the state in which it was collected, to avoid loss or degeneration. In contrast, *in situ* conservation is meant to maintain a living and everchanging system, thus allowing for both loss and addition of elements of the agroecosystem. Just as the conservation of natural habitats and wild species must be ecologically dynamic, we must accept that the *in situ* conservation of crops would fail and collapse if it attempted to stop change or to preserve an agroecosystem in a particular state. Sources of change that can be expected and must be tolerated include the introduction of new crops and crop varieties; exchange of varieties between farmers and localities; the use of inputs to improve the productivity of land and labor, such as fertilizers and pesticides; and commercialization.

The goal of *in situ* conservation is to encourage farmers to continue to select and manage local crop populations. These embody not only diverse alleles and genotypes but also evolutionary processes such as gene flow between different populations and local knowledge systems such as folk taxonomies and information about selection for heterogeneous environments. The primary method for achieving this goal is to increase the value of local and diverse crop populations to farmers who might otherwise stop growing them. The objective here is to raise the value local and traditional crops so that it approximates the social value of genetic. How agricultural agencies and organizations can support *in situ* conservation will be described in the third section of this chapter and elsewhere in the book. Practices that would be detrimental to *in situ* conservation and should thus be discouraged include proscriptions on using particular technology or crops and obligations that bind agricultural credit and other support to the use of particular technologies or crops. An important corollary is that *in situ* conservation will not succeed through administrative coercion.

The dynamic aspect of *in situ* conservation is one of its most difficult attributes for planning and evaluation. Rather than presenting an easily quantified and non-moving target, such as the number of alleles or genotypes in a collection, *in situ* conservation concerns ecological relationships, knowledge, and cultural practices — elements that are difficult to quantify and likely to change over time. The success of *in situ* conservation cannot be judged only by the number of alleles or genotypes preserved. It might also be measured by the number of farmers within a target area or group who maintain local crop populations and manage those populations according to local criteria and practices. Alternatively, the success of *in situ* conservation might be measured by the use of local germplasm in breeding programs that result in new crops but do not replace the crop population of a region. Yet another measure might be the exchange and flow of farmer varieties within and among different communities.

Generating new resources

The second reason for promoting *in situ* conservation is that gene bank collections fail to capture genetic diversity and new resources that are generated after the collection has occurred. The fact that different sampling procedures have been used and documentation is poor in many collections (Frankel et al. 1995) suggest the possibility that much diversity remains uncollected. Estimates of the amount of possible landraces now collected indicate that most diversity of some crops has been captured in gene banks (Plucknett et al. 1987). Such estimates, however, are quickly rendered obsolete by continued crop evolution. Moreover, these estimates are derived from a consensus among scientists rather than from a thorough analysis of genes in the bank and genes in farmers' fields. New resources become available because of a variety of mechanisms — mutation; recombination; gene flow between wild, weedy, and cultivated populations; somatic variation; and exchange from outside the collection region.

Backup to gene bank collection

In situ conservation of crop resources has been criticized because of its potential vulnerability to technological innovation and diffusion, economic and political change, and environmental factors (e.g., Hammer 1996; Zeven 1996). Unfortunately, all forms of conservation are vulnerable, and *ex situ* methods are subject to numerous risk factors — genetic drift within collections, loss of seed viability, equipment failure, security problems, and economic instability. Gene banks, like all human institutions, depend on volatile public and political support. Even large and prestigious institutions may suffer sudden reversals of fortune, endangering their collections. A number of observers point out that gene banks are usually inadequately funded, so that storage and regeneration facilities are limited, evaluation is partial, and equipment is obsolete or not adequately backed up. While the purpose of *in situ* conservation is not to preserve alleles and/or genotypes *per se*, regions where successful on-farm maintenance of genetic diversity occurs provide potential stores for re-collection of genetic resources.

Nevertheless, complementarity between *in situ* and *ex situ* conservation goes beyond a simple backup role for the former. *Ex situ* collections and their associated crop improvement programs give rise to one type of diversity, with selection directed by crop science and commercial and public breeding interests. *In situ* conservation can theoretically generate far more diversity and, perhaps more importantly, selection is directed by farmers in response to local needs and conditions. Thus, *in situ* maintenance of diversity might well produce crops that are adapted to conditions that are not included in the programs of commercial and public crop breeders. New crops resulting from *in situ* maintenance might be especially important to particular farm groups and areas, for instance, in marginal areas such as rain-fed conditions or uplands. *In situ* conservation thus complements *ex situ* maintenance by preserving a stock of genetic diversity that is relevant to farm sectors not reached by commercial and public breeding programs. In this way, *in situ* conservation helps to maintain not only key elements that are missed by *ex situ* methods but also aids in generating new material for areas that are often bypassed by crop improvement programs connected to *ex situ* facilities.

Laboratories for agricultural research

In situ conservation areas are important laboratories for two types of agricultural research. First, the understanding of crop evolutionary processes, such as gene flow between wild and cultivated plants, is best carried out in centers of crop origins, diversity, and evolution. This research is critical not only to the study of domestication and crop evolution but also to identifying new sources of genetic material that has undergone natural and artificial selection. Second, agricultural science has become increasingly aware of the importance of broad ecological processes in the design of technology for sustainable production. Genetic diversity is usually seen as a key component of sustainable technology, to manage risk and reduce reliance on chemical inputs. Genetically diverse agroecosystems that harbor evolutionary processes such as gene flow between wild relatives and cultivated species, adaptation to coevolved pests and pathogens, and traditional knowledge systems and farmer selection offer a unique field laboratory to design and evaluate sustainable technologies.

Convention on Biological Diversity mandate

Finally, the Convention of Biological Diversity provides strong justification for sponsoring *in situ* conservation. This convention, originally negotiated in 1992 and ratified by over 160 countries, specifically includes crop genetic resources and indigenous knowledge as items that require *in situ* conservation. Article 2, in defining the use of terms on the convention, includes domesticated or cultivated species as part of biological diversity and genetic resources. Article 8 addresses *in situ* conservation and, within the article, 8(j) identifies "Knowledge, innovations and practices of indigenous and local communities embodying traditional lifestyles relevant for the conservation and sustainable use of biological diversity ..." (Convention on Biological Diversity 1994:9). The Global Environmental Facility (GEF) of the World Bank, UNDP, and UNEP is the interim funding mechanism of the Convention of Biological Diversity. From its beginnings, the GEF has funded projects dealing with the *in situ* conservation of crop genetic resources, including wild crop relatives, in Turkey, Ethiopia, Peru, Lebanon, and Jordan (International Institute for Sustainable Development). Similar projects are under active preparation and review, for example, for Peru and for countries of the "Fertile Crescent" of the Near East.

The scope of in situ conservation

The need for conservation arises because of two fundamental changes in farming systems within regions of crop origins and diversity: (1) the integration of local systems into larger socioeconomic (e.g., market, political, cultural) and technological (e.g., information, inputs) systems, and (2) the growth of population both at the local level and above to magnitudes far in excess of any previous level. The predicted increase of the world's population to 8 billion people by 2025 (Harris 1996) will require developing nations to double yields over present levels, and the means for achieving this are rarely available locally. Socioeconomic integration and population growth represent a sort of continental drift for agricultural evolution, inducing changes that are long term and irresistible.

The production increases required to meet expected population thus inevitably result in the direct competition between local and exotic knowledge, inputs, and crops. Previously, this competition has provoked genetic erosion or the loss of genetic variability in crop populations. The magnitude of the forces behind genetic erosion suggests that it is not a process that can or should be reversed on a wide scale. *In situ* conservation is very likely to be frustrated and fail if it sets as a goal the reversal of the historic and universal trends of integration and economic and technological transformation that have caused genetic erosion in the first place.

Rather, the purpose of *in situ* conservation programs and projects is to conserve specific agroecological, cultural, and biological processes in specific localities so that the historic processes and ecological relationships of crop evolution remain viable therein. In other words, *in situ* conservation is not a sector-wide strategy for a nation's agriculture but one targeted to a few locations. On-farm conservation is not meant as an alternative to agricultural modernization nor is it appropriate to all farmers.

In situ conservation projects imply the selection of specific areas and groups of farmers as participants. While the selection process necessarily involves some centralized decision making, determining size and location of participation for on-farm conservation also requires a high degree of decentralization and exchange between scientists, government officials, and farmers. Possibly, participation in *in situ* conservation projects can be driven by farmer interest rather than by area or location, especially in areas where the ecological and crop information is lacking. Determining the scope of *in situ* conservation requires us to address several criteria: the crop species to be conserved, the physical size of the conservation program, the location and distribution of conservation target areas, and the number of farms and communities that are included. Answering the questions about scope, therefore, involves both biological sciences (e.g., genetics, ecology, population biology) and social sciences (e.g., anthropology, economics, geography). Besides daunting technical issues in determining the scope of *in situ* conservation, financial, institutional, and political factors are also likely to have weight.

Determining the physical size, location of areas, and numbers of farmers of an *in situ* conservation program requires analysis in both the biological and social sciences, and the integration of these fields. One reason for focusing on a single crop species, rather than a complex of crops in a single farming system, is to make this research more feasible and tractable. The purpose of *in situ* conservation, however, is to maintain ecological relationships within centers of crop genetic diversity — within crop populations, between crops and other populations (e.g., wild and weedy relatives, pests and pathogens), and between farming communities and their local crop populations. Because *in situ* conservation concerns itself not only with crop diversity but also with ecological relationships and human factors, determining the scope depends on both population criteria and on human ecological criteria.

Population criteria that are germane to planning *in situ* conservation are similar to sampling design issues discussed by population biologists (e.g., Marshall and Brown 1975) — the measurement of variation and the number of populations to be sampled. Two critical population parameters are identified by Marshall and Brown (1975): (1) the extent of genetic divergence among populations and (2) the level of genetic variation of a population. Like collectors for *ex situ* collections, planners for *in situ* conservation will be concerned with these parameters in selecting target areas. Marshall and Brown (1975) describe divergence among populations according to frequency and distribution of alleles, leading to four different types of alleles, portrayed in Figure 1.1. The critical target for collection is alleles that are locally common. Populations with locally common alleles are thus primary targets for collection and conservation. Common, widespread alleles are likely to be found wherever a crop is grown and rare alleles are hard to capture, given the limits of collecting. These same guidelines form the first criteria for determining the number of populations that should be targeted for *in situ* conservation. Surveying national collections for distribution and frequency of alleles of target species will be the most direct method of determining the number and location of populations to be considered for *in situ* conservation.

Distribution

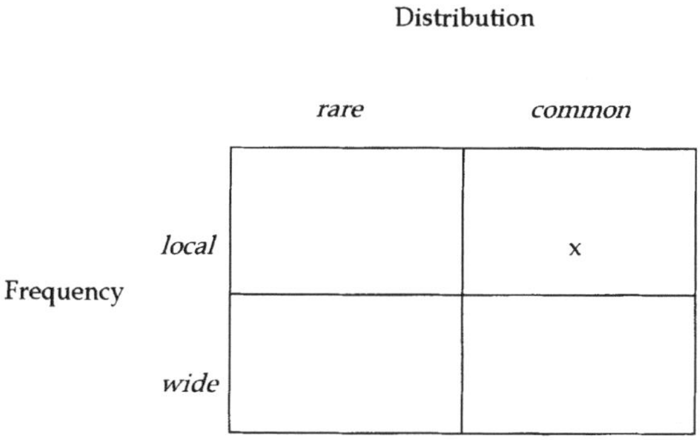

Figure 1.1   Genetic divergence among populations.

The number of ecological criteria is potentially very large and not as easily ranked as the population criteria discussed above. Limited research on these criteria in relation to crop diversity poses immediate problems. Three ecological criteria are identified as being critical to crop diversity and evolution — the presence of wild crop relatives, environmental heterogeneity, and seasonality (Hawkes 1983). Environmental heterogeneity is indicated by such variables as altitude variation and/or a diversity of soils and vegetation biomes within the sample region. Environmental heterogeneity often implies mountainous terrain, a fact that is reinforced by the location of many crops' centers of diversity in such terrain.

Social criteria are equally as important as population and ecological criteria, but they are also numerous and difficult to assess because of their variability and the possibility of rapid social and cultural change. Nevertheless, social scientists have stressed a number of criteria that are associated with the maintenance of crop diversity. Cultural autonomy in terms of local language emphasis and economic autonomy in terms of orientation toward subsistence production are often cited (e.g., Brush et al. 1992; Zimmerer 1996). Subsistence orientation is also expressed in such variables as commercialization, use of purchased inputs, amount of off-farm employment, distance from markets, and access to agricultural extension services.

The population, ecological, and social criteria discussed above can be put into a single matrix that is helpful in determining the size and location of *in situ* conservation areas, shown in Figure 1.2. Here, ecological criteria are expressed as complexity. Places where altitudes, soils, and biomes are varied with seasonality and the presence of wild crop relatives would be judged maximally complex, while locations without these would be classed as having limited complexity. Likewise, cultural autonomy and subsistence orientation can be expressed as local vs. non-local social integration, an idea that is described as level of sociocultural integration by anthropologists (Steward 1955).

		Sociocultural Integration	
		local	*non-local*
Ecological Complexity	*limited*		
	maximum	x	

Figure 1.2 Ecological and social criteria for selecting *in situ* locations.

As indicated in Figure 1.2, the selection of locations using ecological and social criteria draws our attention to one type of location, with local sociocultural integration and ecological complexity. These selection criteria can then be weighed against two other criteria that are essential in identifying sites for *in situ* conservation: crop population criteria and logistical criteria. The selection of regions for an on-farm conservation program might begin with consideration of the population, ecological, and social criteria listed in Figures 1.1 and 1.2 and with consideration of the logistics of the regions. Logistical criteria — physical and social access to the farm region — are also necessary for site selection and scope of on-farm conservation projects. Figures 1.1 and 1.2 can also be used for selecting among sites that are logistically equivalent.

Promoting on-farm conservation

As noted above, the goal of *in situ* conservation is to encourage farmers to continue to select and manage local crop populations, and one method for reaching this goal is to increase the value of local and diverse crop populations to farmers who might otherwise stop growing them. To increase the value of local crop populations and management practices, conservationists must understand the different values that local crops hold for farmers as well as the ways in which changing social and technological conditions will affect those values.

Valuation

Value is difficult to assess because of its inherent subjectivity and because different types of values exist. Value of particular local cultivars is most easily understood when viewing the selection of individual varieties, although crop diversity may be an object in itself. For instance, research on potato landraces in Peru shows that Andean culture values diversity as such and partly explains why so many potato varieties are kept in a farming system where environmental and agronomic factors do not explain diversity (Brush 1992; Zimmerer 1996). It is important to stress that farmers themselves may not value crop genetic resources directly but rather indirectly, by valuing practical and perhaps aesthetic attributes of the crop populations which embody crop genetic resources. Three types of value can be distinguished: direct, indirect, and option value. The first type, direct value, is the most obvious and critical to farmer selection, followed by option and indirect values.

Direct values refer to the harvest and use of crop varieties as part of a noncommercial, commercial, and/or industrial process. Direct values for specific varieties or groups of varieties include the agronomic or environmental assets for production as well as consumption benefits. This type of value is most likely to be recognized and articulated by farmers. Examples of production assets provided by diverse varieties in the farm store include the ability to yield well in distinct environments, as defined by soil classes, altitudes, moisture regimes, and pest and disease conditions. Research on variety choice has revealed that farmers maintain local crop varieties in part because they perform better than other varieties in marginal environments (Brush 1995). In addition to yield advantages of local varieties, farmers may also perceive a risk advantage if these varieties are more stable over time than non-local varieties. Diversity itself may provide yield stability and harvest security in the face of pests, diseases, competition, and unfavorable environments (Clawson 1985), but the relationship between diversity and stability is uncertain and cannot be assumed (Goodman 1975; Pimm 1986). Frankel et al. (1995:61) find support for the proposition that heterogeneity *per se* is adaptive in offering resistance to pathogens, although the idea that the components of diverse landraces are interactive and stabilizing is not supported.

Direct values have been cited as the basis for maintaining diverse local crops in a number of case studies. In a thorough analysis of maize landraces in a village in southern Mexico, Bellon (1996) describes five concerns of farmers that account for infraspecific diversity — environmental heterogeneity, pests and pathogens, risk management, culture and ritual, and diet. These concerns vary among farmers and are influenced by such factors as wealth, land and labor resources, and government policies. No single variety can satisfy the concerns of all the farmers in the village, resulting in the maintenance of a complex population of maize landraces, even though modern varieties and commercial inputs are available.

Besides environmental and production concerns, consumption and other uses of crop varieties are direct values that also influence farmers to maintain local varieties. Consumption values may be associated with special qualities that can be found in local crop varieties but not in non-local ones. These qualities include taste, cooking characteristics, or better storage. Secondary products, such as straw for animal feed, is another quality that imbues local varieties with consumption value. Other consumption values that might be derived from local crop varieties include their significance as prestige, ritual, or gift items. Research among potato farmers in Peru revealed that local varieties were prized for their "floury" texture and were important as gifts and as a means to recruit labor (Brush 1992). In fact, local varieties in the central Andean highlands of Peru are referred to as "gift potatoes" (*papas de regalo*). The predominant role of potatoes in the diet of these farmers is reflected in social and cultural embellishments that reinforce the selection of local and diverse varieties.

Indirect values refer to environmental services rendered by crop varieties and benefits that result from biological resources without depending on harvest and consumption. For *in situ* conservation, the most important asset of local crop varieties is their indirect value in maintaining crop evolutionary relationships. However, these relationships may not be understood or observable to the farmers who maintain local crop populations. Diversity in one crop may, for instance, strengthen polycropping, the cultivation of other crops simultaneously and in the same field. Diversity in one crop may, therefore, add to diversity of others. An example is the association of traditional types of maize in Mexican agriculture with beans and squash, in the *milpa* system. Beans are beneficial because of their association with nitrogen-fixing bacteria, and both beans and squash add important nutrients to the diet, increasing the overall return to land and labor. Modern maize varieties in Mexico tend to be grown in monocropped fields.

Option values derive from future use of a resource (Krutilla 1967). The option value of a crop variety may be expressed as the desire to bequeath a family or cultural patrimony to future generations or as the potential of a variety to meet future demands or conditions of production. The idea that local crop varieties represent a bequest to future generations has been noted in places where particular varieties are associated with a family lineage (Sutlive 1974) or as an expression of the farming knowledge and skill of parents (Zimmerer 1996). The potential value of varieties is likely to be recognized by farmers in areas where seed is produced and exported to other regions. Seed production areas have been described for potato farming in Peru (Brush et al. 1981), in Mexican maize agriculture (Louette et al. 1997), and in Thai rice agriculture (Dennis 1987). Such seed producing areas would appear to be excellent choices for *in situ* conservation programs. Areas that experience a high turnover rate of crop varieties might be appropriate for *in situ* conservation in connection with seed producing areas. Thai rice farmers, for example, were found to acquire and discard local varieties at very high rates. Using variety names as the basis for turnover, Dennis (1987) found that between 1950 and 1982, only 22 out of 122 rice varieties remained in the inventory of farmers in his study area. Selecting both seed producing and seed receiving areas puts both the generation of diversity and its selection into an *in situ* conservation program.

In situ conservation programs can most easily address the first (direct) and third (option) types of value. A variety of tools are available to increase the value of local crop varieties, and these tools can be roughly classified into two different categories — market methods and non-market methods. These methods are not unique to valuation of crop genetic resources but are drawn from other agricultural development programs and adapted to on-farm conservation.

Market methods

Two general market methods are available for increasing the direct and option values of local crops and management. One depends on developing market channels for local produce to increase the value of crops that have genetic resources. The other relies on legal mechanisms for restricting the supply of genetic resources, thereby raising their value for sale as genetic resources. The first method is a form of "green marketing," similar to programs to develop products and markets from biologically important areas that depend on sustainable harvest rather than ecological conversion. The second method implies the creation or use of intellectual property for genetic resources that are found in farmers' fields and farm stores.

The first approach to marketing may be useful in the marketing of landraces with crop genetic resources as consumer goods rather than as germplasm. Most farmers in the world are now involved to some degree in markets. Local and regional markets may be predominant to most farmers who maintain crop genetic resources, but these markets are usually linked to larger national and urban markets. It is not at all unusual to observe small amounts of local or traditional crops or their products for sale in large urban markets, often in the informal sector of those markets. For example, research on the production and use of diverse potato landraces in Peru found that while local varieties were grown primarily for home consumption, a regular and appreciable amount of the production of local varieties was marketed (Brush 1992). Selected local potato varieties are also grown as commercial crops in fields of single varieties. These selected varieties are in much demand in urban markets, where consumers are willing to pay higher prices for these types than the higher yielding modern varieties. Urban demand contributes to the value Peruvian potato farmers in the highlands attach to potato landraces, helping to perpetuate the Andean *chacra* system that generates potato resources.

Marketing of products for environmental, humanitarian, and other social causes has proved to be a successful method to support public causes. "Green marketing" has been particularly developed (Wasik 1996; Peattie 1995). This type of marketing identifies the product with particular qualities of consumer interest, such as "organic" or "dioxin free," or with beneficial characteristics, such as "biodegradable" (Wasik 1996). Regulated labeling guarantees the products of this type of marketing in some places and for certain qualities. In many countries, there are systems to certify that organically produced food and products meet state standards to be labeled "certified organic." The growth of both national and international trade associated with environmental or social causes has been sustained and robust for a decade or more, showing that consumers are willing to support these causes.

A green marketing approach has not been directly attempted for increasing the value of landraces with crop genetic resource properties, but the strength of this market for other products, interest in preserving local culture, and the existence of a limited trade in landraces are positive indications. A stronger market for landraces can enhance their value without the cumbersome legal framework required of contracting and appellation. Developing the local and national market for landraces may be accomplished in several ways. The identification of special "niche markets" where landraces are in demand and information on the marketing channels that bring landrace produce to market can suggest bottlenecks and constraints to the market, for example, the lack of adequate storage and transportation facilities or information or inadequate supplies of landraces for market. Supplies may be inadequate because of the lack of credit or other inputs. Market constraints might be overcome directly as part of *in situ* conservation projects, for example, through supporting facility construction, promotional campaigns for landrace products, and helping to increase production of landraces for market. *In situ* conservation programs may also work with private manufacturers of food products to incorporate landraces and promote the products as part of national and international effort to meet conservation and development goals.

An example of a successful green marketing program for promoting *in situ* conservation has been established in the U.S. to maintain and utilize ancestral maize by Cherokee farmers in North Carolina (Brown and Robinson 1992). Maize landraces of the Cherokee had been greatly reduced by the diffusion of hybrid maize and contamination with commercial varieties.

Maize scientists located remnant populations on a few farms and in collections. Through controlled genetic matings, Brown and Robinson (1992) were able to reestablish pure Indian flour maize, which is now grown and marketed by the Cherokee Boys Club in North Carolina.

One approach that might be applied to increasing the value of local crops through the market is the appellation or certification system of restricted labeling. Appellation relies on legal enforcement, on market demand, and willingness to pay additional costs for the guarantee of the appellation. The appellation system is well developed for high quality food products in Europe, for example, for wine, cheeses, and meats (Bérard and Marchenay 1996). To date, the appellation system is based on geographic location or manufacture, to ensure quality and authenticity. In at least one instance, however, a certification system has been developed to guarantee the origin of plants (Meilleur 1996).

It is also possible that an appellation or certification can be attached to genetic resources for the purpose of financing *in situ* conservation. Crop genetic resources with diversity and evolutionary potential might be covered by an appellation system that could designate crop genetic resources from particular regions. This system might succeed if seed companies or other "users" of genetic resources accepted the social obligation to underwrite *in situ* conservation as they seek to acquire crop genetic resources. Nevertheless, the limits to a market for crop genetic resources which affect contracting for biological prospecting are also likely to be detrimental to a market for crop genetic resource appellation. There is currently no national or international market for crop genetic resources, and the crop breeding industry is not likely to generate a market. Contracting for genetic resources and labeling for appellation are, therefore, unlikely to be the first or most useful market methods to increase direct and option values of local crops. The transaction costs of establishing and maintaining legal mechanisms to increase the value of local crops are probably high and above the potential market price of crop genetic resources.

The second general market method for increasing the value of local and traditional crops involves the direct sale of genetic resources, either under contract or as intellectual property. Contracting for "biological prospecting" has been used in a limited way for pharmaceutical products (Reid et al. 1993; King et al. 1996), and similar contracts for crop genetic resources from functioning agricultural systems may be possible. Besides the legal framework of contracting, these agreements also involve intellectual property and sharing royalties that derive from the resources (Gollin and Laird 1996). The value of a contract to a farming community is commensurate with the exclusivity that a community can claim for its genetic resources. Given the open exchange of seeds that pervades agriculture, efforts to claim and defend exclusive ownership over genetic resources will be extremely difficult and therefore expensive. The likelihood of contracting for crop genetic resources is also greatly reduced by the existence of large public collections which provide crop germplasm without charge and with information that is usually not available directly from farmers. Moreover, many seed companies rely on their own working collections and on breeding material from other public and private agencies. The demand for germplasm from farmers' fields is, therefore, likely to be small and unable to generate sufficient commercial demand to finance conservation (Brush 1996). Surveys of breeders and breeding programs suggest that this is generally the case (Goodman 1985; Peeters and Galeway 1988; Marshall 1989; Rejesus et al. 1996).

Non-market methods

The direct and option value of diverse local crops can also be increased by methods that do not rely on the market. Two non-market approaches have been developed for promoting *in situ* conservation: (1) educational or promotional campaigns and (2) increased use of local crop resources and farmer participation in crop breeding and improvement programs.

In the Peruvian Andes, a seat of great crop diversity, governmental and non-governmental organizations have sponsored a visible and popular system of "diversity fairs," which bolster the value of local crops and promote conservation. Farmers from different villages assemble and display the diversity of local crops grown in their villages. Tubers are especially prominent — potatoes (*Solanum* spp.), ocas (*Oxalis tuberosa*), mashua or añu (*Trapaeolum tubersum*), olluco (*Ullucus tubersus*), achira (*Canna edulis*), arracacha (*Arracacia xanthorriza*), and yacón (*Polymnia sonchifolia*); but native and introduced cereals, legumes, fruits, and vegetables are also displayed. Natural pride in showing and interest in observing diversity are sufficient to stimulate enthusiasm for the diversity fair. The promoters have also found that public gifts, such as materials for school construction or repair, add to the effectiveness of the diversity fair. Diversity fairs are part of a broader set of activities to promote and enhance the production of local crop varieties, including information on cultural practices to improve production and the development of markets for selected local potato varieties.

Perhaps the most important strategy for increasing the value of local crops is to use them as the basis for crop improvement programs, especially with the participation of farmers who will use the results. This approach is referred to as "Participatory Plant Breeding." Participatory plant breeding is defined as the formalized cooperation between farmers and plant breeders in such activities as identifying crop improvement needs and priorities, selecting varieties, and evaluating varieties (Eyzaguirre and Iwanaga 1996). The *in situ* conservation aspect of participatory plant breeding is to offer farmers a viable alternative to using exotic crops and varieties in their quest to increase production or income. Development of local varieties and populations through procedures such as mass selection may be especially suitable for marginal environments where conventional breeding has had limited success. Participatory plant breeding thus provides not only a context for *in situ* conservation but also one to work in environments where normal crop improvement has been frustrated.

Two distinct levels of participatory plant breeding are distinguished in the literature. Participatory plant breeding is defined as the selection by farmers of "genotypes from genetically variable, segregating material," while participatory varietal selection is the "selection by farmers of nonsegregating, characterized products from plant breeding programs" (Whitcombe and Joshi 1996). Participatory plant breeding thus means farmer participation is selection at the F_2 level and above, while participatory varietal selection means farmer participation at the F_5 level and above. Actually, most proponents of participatory plant breeding refer to what Whitcombe and Joshi (1996) define as participatory varietal selection (e.g., Berg 1996; Eyzaguirre and Iwanaga 1996; Weltzien et al. 1996). A smaller group proposes participatory plant breeding rather than participatory varietal selection, involving cooperation between farmers and breeders from the early stages of selecting segregating populations (e.g., Sperling et al. 1993; Ashby and Sperling 1995; Ashby et al. 1996).

Both participatory plant breeding and participatory varietal selection are likely to have negative impacts on diversity of landraces, because both methods are intended to change local crop population structure to make it higher yielding. Nevertheless, participatory plant breeding is likely to be more beneficial for conservation goals because it works with variable, segregating material that is derived from or similar to material already in the local farming system. Participatory varietal selection, on the other hand, is likely to be negative for conservation goals because it is based on replacement of local populations with new and less variable ones from breeding programs.

Institutional issues

This chapter has affirmed that *in situ* conservation is complementary to *ex situ* conservation and that its scope should be modest and specific rather than system-wide for agriculture in less developed countries. The remaining institutional issue is legal — the ownership of and compensation for genetic resources, local knowledge, and new plant varieties. A longstanding debate about "Farmers' Rights" contrasts the interests of industrial countries that use genetic resources against the interests of non-industrial countries that produce them. Industrial countries are concerned with access to genetic resources and with protecting intellectual property that they have recognized. Non-industrial countries are interested in sharing the financial and technological benefits derived from using genetic resources. Conflict between these two parties surrounds the granting of intellectual property, compensation for resources normally considered to be public goods, and the ownership of resources already collected. Because *in situ* conservation provides a pool of genetic resources for future collection, these conflicts necessarily arise.

Biological resources and indigenous knowledge have conventionally been collected under the principle of common heritage (Brush 1996; Fowler and Mooney 1990). This principle defines genetic resources as public domain goods, like products of nature, scientific theory, and folk knowledge. Public goods are defined by the quality of non-competitiveness — one person's use of elements within the public domain does not deprive use to others. Public goods, however, can be removed from the public domain through intellectual property.

Although public goods are "free," they may involve costs — for example, the cost of keeping air and water clean. One problem with public goods is that the costs of maintaining them are difficult to calculate because market pricing is not possible without private or exclusive use provisions. Another problem is that individuals who are directly involved in producing or maintaining public goods may not be fairly compensated. Thus, farmers who soundly manage hillside fields and pastures may not be rewarded for the clear water flowing off their farms, nor are they compensated for protecting water supplies. Likewise, farmers who produce crop genetic resources are not compensated for their costs in maintaining them.

The difference between the social and private values of public goods is problematic because individuals may not have sufficient personal incentive to maintain socially valuable goods, such as clean water or genetic resources. Theoretically, the deterioration of public goods, such as water pollution or soil erosion, is attributable to the imbalance between private and social values of public goods (Sedjo 1992; Vogel 1994). In the case of crop genetic resources, the loss of genetic diversity, or genetic erosion, is analogous to the loss of topsoil from common pastures. In each of these examples, farmers receive little reward for producing socially beneficial goods or compensation for the costs of producing or maintaining those goods.

The genetic resources of crop in centers of diversity illustrate the problems of estimating and rectifying public and private values. Maize farmers in Iowa or Africa receive part of the public value of maize landraces cultivated by Zapotec farmers in Mexico, yet Zapotec farmers bear an uncompensated private cost for keeping maize landraces, the cost of forgoing alternatives to plant modern maize varieties or other crops, or to leave agriculture altogether. Moreover, the fact that seed companies and others in industrial countries can claim exclusive ownership of the results of their use of genetic resource seems unfair, especially because farmers provided the essential resources without compensation.

This chapter has argued that contracts for crop genetic resources are unlikely to generate financial rewards because of the large supply and small demand for them. These same financial constraints are likely to confront efforts by non-industrial countries and/or farmers to gain financially from a novel form of intellectual property, such as Farmers' Rights. Moreover, there appears to be little opportunity to create a novel form of compensation to farmers. The strength of the movements to extend intellectual property to new geographic areas and to include plant materials has been reinforced and clearly demonstrated by the last (Uruguay Round) General Agreement on Tariffs and Trade (GATT), with the agreement on Trade Related Aspects Intellectual Property Rights (TRIPs). Nations that do not have systems for intellectual property protection of plant varieties are obligated by GATT and TRIPs to create them, and both the prevailing legal models and political pressure favor conventional Breeders' Rights rather than Farmers' Rights. The international system of intellectual property appears to be moving in the direction of creating more stringent and uniform standards, a direction away from defining novel rights such as Farmers' Rights.

Nevertheless, the issues of ownership of crop genetic resources and compensation to farmers for maintaining and providing crop genetic resources are widely discussed among donors, non-governmental organizations, conservation agencies, and governments. The issues of ownership and compensation are necessarily addressed by national legislation and policy processes, including international treaties and trade, that extend far beyond the purview of *in situ* conservation. The planning and implementation of *in situ* conservation must, however, be cognizant of these policy issues and can address them in a limited way by recognizing the contributions of farmers in providing genetic resources and affirming the need to include these farmers as active participants in the worldwide conservation effort.

Conclusion

This chapter has addressed four major questions that confront any program for *in situ* conservation of crop genetic resources. There are no general or definitive answers to these questions; rather, answers must be context specific, to the country, region, crop, and farming system. The approach here has been to identify guideposts for answering these four questions at the national and local levels. The past decade has seen a burgeoning of research in the ecology and biogeography of crop genetic resources in several countries and for different crops. This research has prepared us to answer these questions, but the novelty of this area of research, its interdisciplinary nature, and the complexity of the topic make it difficult to find ready answers to the questions posed at the beginning of this chapter. In trying to answer these questions, however, agricultural researchers and their farmer partners have taken the first strides in implementing *in situ* conservation.

References

Altieri, M.A. and L.C. Merrick. 1987. *In situ* conservation of crop genetic resources through maintenance of traditional farming systems, *Economic Botany* 41:86–96.

Ashby, J.A. and L. Sperling. 1995. Institutionalizing participatory, client-driven research and technology development in agriculture, *Development and Change* 26:753–70.

Ashby, J.A., T. Garcia, M. del Pilar Guerrero, C.A. Quiros, J.I. Roa, and J.A. Beltran. 1996. Innovation in the organization of participatory plant breeding. In *Participatory Plant Breeding*, P. Eyzaguirre and M. Iwanaga (eds.). Rome: International Plant Genetics Resources Institute.

Bellon, M.R. 1996. The dynamics of crop infraspecific diversity: A conceptual framework at the farmer level, *Economic Botany* 50:26–39.

Bérard, L. and P. Marchenay. 1996. Tradition, regulation, and intellectual property: Local agricultural products and foodstuffs in France. In *Valuing Local Knowledge: Indigenous People and Intellectual Property Rights*, S. Brush and D. Stabinsky (eds.). Washington, D.C.: Island Press.

Berg, T. 1996. The compatibility of grassroots breeding and modern farming. In *Participatory Plant Breeding*, P. Eyzaguirre and M. Iwanaga (eds.). Rome: IPGRI.

Brown, W.L. and H.F. Robinson. 1992. The status, evolutionary significance and history of eastern Cherokee maize, *Maydica* 37:29–39.

Brush, S.B. 1991. A farmer-based approach to conserving crop germplasm, *Economic Botany* 45:153–161.

Brush, S.B. 1992. Ethnoecology, biodiversity, and modernization in Andean potato agriculture, *Journal of Ethnobiology* 12:161–85.

Brush, S.B. 1995. *In situ* conservation of landraces in centers of crop diversity, *Crop Science* 35:346–354.

Brush, S.B. 1996. Valuing crop genetic resources, *Journal of Environment and Development* 5:418–435.

Brush, S.B., H.J. Carney, and Z. Hauman. 1981. Dynamics of Andean potato agriculture, *Economic Botany* 35:70–88.

Brush, S.B. and E. Meng. 1998. Farmers' valuation and conservation of crop genetic resources, *Genetic Resources and Crop Evolution* 45:139–150.

Brush, S.B., J.E. Taylor, and M.R. Bellon. 1992. Biological diversity and technology adoption in Andean potato agriculture, *Journal of Development Economics* 39:365–387.

Cernea, M. M. (ed.) 1985. *Putting People First: Sociological Variables in Rural Development* (2nd ed.). New York: World Bank by Oxford University Press.

Chambers, R. 1983. *Rural Development: Putting the Last First*. London: Longman.

Clawson, D.L. 1985. Harvest security and intraspecific diversity in traditional tropical agriculture, *Economic Botany* 39:56–67.

Cochrane, W.W. 1993. *The Development of American Agriculture: A Historical Analysis* (2nd ed.). Minneapolis: University of Minnesota Press.

Convention on Biological Diversity. 1994. *Convention on Biological Diversity Text and Annexes*. Geneva: Interim Secretariat for the Convention on Biological Diversity.

Dennis, J.V. 1987. Farmer Management of Rice Variety Diversity in Northern Thailand. Unpublished Ph.D. dissertation, Cornell University. UMI No. 8725764. University Microfilms International, Ann Arbor, MI.

Eyzaguirre, P. and M. Iwanaga (eds.). 1996. *Participatory Plant Breeding. Proceedings of a Workshop on Participatory Plant Breeding*, 26–29 July 1995. Wageningen, the Netherlands. Rome: International Genetic Resources Institute.

Fowler, C. and P. Mooney. 1990. *Shattering: Food, Politics and the Loss of Genetic Diversity*. Tucson: University of Arizona Press.

Frankel, O.H. 1970. Genetic conservation in perspective. In *Genetic Resources in Plants — Their Exploration and Conservation*, O. H. Frankel and E. Bennett (eds.). Oxford: International Biological Programme Handbook No. 11, Blackwell Scientific Publications.

Frankel, O.H., A.H.D. Brown, and J.J. Burdon. 1995. *The Conservation of Plant Biodiversity.* Cambridge: Cambridge University Press.

Gollin, M.A. and S.A. Laird. 1996. Global politics, local actions. The role of national legislation in sustainable biodiversity prospecting, *Boston University Journal of Science & Technology Law* (Lexus, 2 B. U. J. Sci. & Tech. L. 16).

Goodman, D. 1975. The theory of diversity–stability relationships in ecology, *The Quarterly Review of Biology* 50:237–266.

Goodman, M.M. 1985. Exotic maize germplasm, status, prospects, and remedies, *Iowa State Journal of Research* 59:497–527.

Hammer, K., H. Knupffer, L. Xhuveli, and P. Perrino. 1996. Estimating genetic erosion in landraces — two case studies, *Genetic Resources and Crop Evolution* 43:329–336.

Harlan, H.R. and M.L. Martini. 1936. Problems and results of barley breeding. *USDA Yearbook of Agriculture*, pp. 303–346. Washington, D.C.: U.S. Government Printing Office.

Harlan, J.R. 1995. *The Living Fields: Our Agricultural Heritage.* Cambridge: Cambridge University Press.

Harris, D.R. 1989. An evolutionary continuum of people–plant interaction. In *Foraging and Farming: The Evolution of Plant Exploration*, D.R. Harris and G.C. Hillman (eds.). London: Unwin-Hyman.

Harris, D.R. and G.C. Hillman (eds.). 1989. *Foraging and Farming: The Evolution of Plant Exploration.* London: Unwin-Hyman.

Harris, J.M. 1996. World agricultural futures: regional sustainability and ecological limits, *Ecological Economics* 17:95–115.

Hawkes, J.G. 1983. *The Diversity of Crop Plants.* Cambridge, MA: Harvard University Press.

Hayami, Y. and V. Ruttan. 1985. *Agricultural Development: An International Perspective* (2nd ed.). Baltimore: The Johns Hopkins University Press.

King, S.R., T.J. Carlson, and K. Moran. 1996. Biological diversity, indigenous knowledge, drug discovery, and intellectual property rights. In *Valuing Local Knowledge: Indigenous People and Intellectual Property Rights*, S. Brush and D. Stabinsky (eds.). Washington, D.C.: Island Press.

Krutilla, J.V. 1967. Conservation reconsidered, *American Economic Review* 57:777–86.

Louette, D., A. Charrier, and J. Berthaud. 1997. *In situ* conservation of maize in Mexico. Genetic diversity and maize seed management in a traditional community, *Economic Botany* 51:20–39.

Marshall, D.R. 1989. Limitations to the use germplasm collections. In *The Use of Plant Genetic Resources*, A.H.D. Brown, O.H. Frankel, D.R. Marshall, and J.T. Williams (eds.). Cambridge: Cambridge University Press.

Marshall, D.R. and A.H.D. Brown. 1975. Optimum sampling strategies in genetic conservation. In *Crop Genetic Resources for Today and Tomorrow*, O.H. Frankel and J.G. Hawkes (eds.). International Biological Programme 2. Cambridge: Cambridge University Press.

Maxted, N., B. Ford-Lloyd, and J.G. Hawkes. 1997a. *Plant Genetic Conservation: The* in situ *Approach.* London: Chapman & Hall.

Maxted, N., B. Ford-Lloyd, and J.G. Hawkes. 1997b. Complementary conservation strategies. In *Plant Genetic Conservation: The* in situ *Approach*, N. Maxted, B. Ford-Lloyd, and J.G. Hawkes (eds.). London: Chapman & Hall.

Meilleur, B.A. 1996. Selling Hawaiian crop cultivars. In *Valuing Local Knowledge: Indigenous People and Intellectual Property Rights*, S. Brush and D. Stabinsky (eds.). Washington, D.C.: Island Press.

Peattie, K. 1995. *Environmental Marketing Management: Meeting the Green Challenge.* London: Pitman.

Peeters, J.P. and N.W. Galwey. 1988. Germplasm collections and breeding needs in Europe, *Economic Botany* 42:503–521.

Pimm, S.L. 1986. Community stability and structure. In *Conservation Biology,* M.E. Soulé (ed.). Sunderland, MA: Sinauer Associates, Inc.

Plucknett, D.L., N.J.H. Smith, J.T. Williams, and N.M. Anishetty. 1987. *Gene Banks and the World's Food.* Princeton, NJ: Princeton University Press.

Reid, W.V., S. Laird, C. Meyer, R. Gámez, A. Sittenfeld, et al. 1993. *Biodiversity Prospecting, Using Resources for Sustainable Development.* Washington, D.C.: World Resources Institute.

Rejesus, R.M., M. Smale, and V. Van Ginkle. 1996. Wheat breeders' perspectives on genetic diversity and germplasm use: findings from an international survey, *Plant Varieties and Seeds* 9:129–147.

Sedjo, R.A. 1992. Property rights, genetic resources, and biotechnological change, *Journal of Law and Economics* 35:199–213.

Sperling, L., M.E. Loevinsohn, and B. Ntabomvura. 1993. Rethinking the farmer's role in plant breeding: Local bean experts and on-station selection in Rwanda, *Experimental Agriculture* 29:509–519.

Steward, J.H. 1955. *Theory of Culture Change.* Urbana, IL: University of Illinois Press.

Sutlive, V.H. 1978. *The Iban of Sarawak.* Arlington Heights, IL: AHM Publishing Corporation.

Vavilov, N.I. 1926. *Studies on the Origin of Cultivated Plants.* Leningrad: State Press.

Vogel, J.H. 1994. *Genes for Sale, Privatization as a Conservation Policy.* New York: Oxford University Press.

Wasik, J.F. 1996. *Green Marketing and Management: A Global Perspective.* Cambridge, MA: Blackwell Business.

Weltzein, R.E., M.L. Whitaker, and M.M. Anders. 1996. Farmer participation in pearl millet breeding for marginal environments. In *Participatory Plant Breeding,* P. Eyzaguirre and M. Iwanaga (eds.). Rome: IPGRI.

Whitcombe, J. and A. Joshi. 1996. Farmer participatory approaches for varietal breeding and selection and linkages with the formal sector. In *Participatory Plant Breeding.* P. Eyzaguirre and M. Iwanaga (eds.). Rome: IPGRI.

Wilkes, H.G. 1995. The ethnobotany of artificial selection in seed plant domestication. In *Ethnobotany: Evolution of a Discipline,* R. E. Shultes and S. von Reis (eds.). Portland, OR: Discorides Press.

Zeven, A.C. 1996. Results of activities to maintain landraces and other material in some European countries *in situ* before 1945 and what we may learn from them, *Genetic Resources and Crop Evolution* 43:337–341.

Zimmerer, K.S. 1996. *Changing Fortunes: Biodivesity and Peasant Livelihood in the Andes.* Berkeley: University of California Press.

Section II

Population biology
and social science

chapter two

The genetic structure of crop landraces and the challenge to conserve them in situ on farms

Anthony H. D. Brown

Introduction

In situ conservation of agricultural biodiversity is the maintenance of the diversity present in and among populations of the many species used directly in agriculture or used as sources of genes in the habitats where such diversity arose and continues to grow. Broadly, the species targets of on-farm genetic conservation include cultivated crop, forage, and agroforestry species, as well as the wild relatives of cultivated species that may be growing in adjacent disturbed sites. This chapter, however, will discuss primarily on-farm conservation of cultivated species as distinct from spontaneously growing populations. The main targets are the landraces or heterogeneous crop populations that humans deliberately cultivate: those that are not the products of modern plant breeding or subject to purifying selection. Planning for the conservation of this kind of biodiversity *in situ* is novel and contentious. The conventionally accepted role for *in situ* strategies in genetic conservation has been in the conserving of wild species. In contrast, *ex situ* collections are the predominant strategy for conserving the genetic variation of cultivated species (Frankel and Soulé 1981; Marshall 1989).

For the on-farm conservation of domesticated species, the traditional cultures and cropping systems that grow and use such populations are fundamental aspects of the habitats to which they are adapted. The systems shape their present genetic structure and determine the changes within landrace populations. Hence, farmers are crucial partners in the process of

in situ conservation. "*In situ* conservation specifically refers to the maintenance of variable populations in their natural or farming environment,
within the community of which they form a part, allowing the natural
processes of evolution to take place" (Qualset et al. 1997:165).

Impetus for the plant genetic resources community to turn its attention
to *in situ* conservation of cultivated populations on farms has arisen from
diverse sources. Perhaps most evocative have been recent studies of crop
diversity that indicate much diversity still persists on farms in regions known
as centers of diversity, despite the advent of modern cultivars of crops to
those areas (Brush 1995). The Rio Convention on Biodiversity (United
Nations 1992) has underlined the challenge for each country to husband its
genetic resources so that countries are not solely dependent on a few *ex situ*
collections or on foreign public or private breeding programs.

There is now widespread recognition of the need to plan for *in situ*
conservation to continue and indeed to improve its capacity to maintain
genetic diversity as an adjunct to conservation in *ex situ* collections. The
need for efficiency is likely to increase as the areas currently devoted to
traditional varieties are subject to increasing pressures to convert to
advanced cultivars. Equally, there is recognition that the scientific basis
and the optimal procedures for on-farm conservation are lacking. As a basis
for guiding the supporting research in population biology for on-farm
conservation, this chapter reviews recent studies of the genetic structure
of landraces of crops in relation to the special advantages attributed to *in
situ* conservation of these genetic resources.

Postulated advantages of in situ *conservation of landraces*

The *in situ* approach to conserving landraces is reputed in the literature to
hold several important advantages over *ex situ* strategies. These advantages
form a convenient framework for developing a research agenda, and for
optimizing methods. In seeking to strengthen the scientific basis of *in situ*
conservation, we should critically assess the nature and extent of evidence
that currently support these advantages. They form the hypotheses to which
research should be directed. The advantages are:

1. **Conservation of indigenous knowledge** — Farmers are central participants in the *in situ* effort. The conservation of crop genetic diversity
   on farms retains the diversity within its proper ethnobotanical context. At the same time, on-farm conservation maintains indigenous
   knowledge about the farming systems and agricultural practices that
   retain diversity and knowledge about direct uses of that diversity.
   Unfortunately, there is relatively little information about the dynamics
   of this kind of indigenous knowledge.
2. **Conservation linked with use** — On-farm conservation is closely
   connected with use directly by the farmer for food or sale. Other uses
   of such populations, either as the source of elite sub-lines or as donors

of characters in advanced breeding programs, require development and monitoring. The use of genetic resources conserved in collections *ex situ* has been a matter of concern, particularly if collections are underutilized and vulnerable to loss of support (Brown et al. 1989).

3. **Allelic richness and genotypic diversity** — On-farm populations have the capacity to support a much greater number of rare alleles and of different (multilocus) genotypes than accessions in gene banks (Marshall 1989). For this feature to apply, large numbers of individual plants with autonomous ancestry must be grown over significant areas. This suggests the need for and importance of measures of the area devoted to landraces, the numbers of populations and their sizes, and the genetic diversity for marker loci, disease resistance, and morphological traits.

4. **Special adaptations** — The *in situ* strategy conserves a unique constellation of germplasm, particularly for marginal or stress environments. This provokes the question of how populations on the farm relate to material already in *ex situ* collections generally, and stored accessions from that specific region. An important indicator of the distinct value of *in situ* populations is the relative ease with which new cultivars are extracted simply and directly as controlled selections out of the variable unimproved populations.

5. **Localized divergence** — The *in situ* strategy conserves genetic variation on a relatively fine spatial scale, in theory down to the individual field. This capacity, however, raises the question of what scale of divergence is reached in practice. Further, the long-term and broad significance of fine-scale differentiation is open to question. Is it important to maintain separate populations aimed at conserving fine-scale differences separately? It is unlikely that such subtle differences will have any use in breeding programs.

6. **Diversity to meet temporal environmental variation** — Diversity itself confers long-term population fitness because it helps populations to cope with variable environments. Landrace populations of crops have survived centuries of selection for reliable production in subsistence agriculture, yielding a definite, known but probably limited benefit to the farmers that grow them (Frankel et al. 1995). Presumably they are selected for resilience and stability though modest productivity, rather than outstanding productivity in the more favorable years.

7. **Continuing crop evolutionary processes** — The *in situ* strategy conserves the crop evolutionary processes (mutation, migration, recombination, and selection). It provides scope for ongoing evolution, particularly in response to environmental changes and pathogen and pest pressures fluctuating in numbers and genetic composition. The key variables are (i) genetic diversity within populations, which is the essential raw material for evolution; (ii) breeding system variation (such as changes in outcrossing rate); (iii) variation in resistance in

space and time, related to pest pressure and diversity; and (iv) the dynamics of seed systems, persistence, and migration.

8. **Avoidance of regeneration** — Regeneration of *ex situ* collections is currently considered a serious and enormous challenge (Brown et al. 1997). Viability is inevitably lost at rates depending on the resources for and the management regimes of such collections. The task is to regenerate accessions without incurring genetic drift (from small samples) or genetic shift (from inadvertent selection in an environment remote from the origin of the accession) (Breese 1989).

9. **Human involvement** — In on-farm conservation, the effort is shared among many players and is thus less dependent on the commitment of one institution or country. However, steady if not rapid rural and social change can occur over wide areas with attendant loss of genetic diversity. Zeven (1996) recounts how prewar recommendations to conserve European landraces *in situ* were entirely disregarded and these populations disappeared. This experience led him to be skeptical of maintenance *in situ* in the longer term, without continuing support and a direct benefit to farmers.

10. **Control and benefit sharing** — Local control of landraces and access to them can ensure that benefits, if any, accrue to the farmers and communities that developed them. This requires workable and not unduly restrictive policies of access.

In summary, Numbers 1 and 2 are comparative advantages that refer to farmers, Numbers 3 to 7 refer to the genetic structure of landraces themselves, and Numbers 8 to 10 refer to management issues and the policy environment. We now turn to evidence and research explicitly aimed at understanding the genetics of landrace populations.

Scientific research issues that underpin on-farm conservation

Worede (1997) among others has pointed out that *in situ* conservation of landraces is already happening. Considerable evidence attests that farmers in various regions of crop diversity [e.g., cereals in Ethiopia; maize and potatoes in Peru; rice in Thailand (Brush 1995)] are growing local diverse varieties, often in small patches amid modern cultivars. This suggests the first important research question is to discover why farmers continue to do so.

Why are traditional varieties being grown without external financial inducements?

Several reasons are likely to apply: advanced varieties may not be available or affordable; advanced varieties may not represent an advance for a particular farmer or meet the farmer's needs reliably; and traditional varieties have cultural or aesthetic appeal or occupy a market niche. More survey evidence, however, is needed to answer this question fully, particularly to gauge how often landraces are grown because of the lack of other varietal options as opposed to more positive reasons such as the filling of special needs.

Qualset et al. (1997) noted that landraces continue to be grown but in shrinking-sized, minority patches. The temporal dynamics of such landrace patches (the extinction–recolonization cycle of fragmentation theory) is as yet little known. Fragmentation arises from the conversion of the land to exotic cultivars, to other cultivated species, to other land uses, or altered agricultural systems. The authors note the likely dynamic factors to include in a study of the retention of landraces at the landscape level are the division of land holdings within families, marginal agricultural conditions associated with hill lands, heterogeneous soils, economic isolation, niche market premiums, cultural values, and specific uses and preference for diversity. The parameters of this area of research suggest the need to draw on both anthropological and genetic expertise and the interaction between them.

To understand the dynamics of local crop diversity in farming systems, we need to relate farmers' decision making to the pool of varieties available for planting. Bellon (1996) outlines a framework to accomplish this. The framework assumes that the farmer has several concerns, including adverse climate, soils, labor or fertilizer shortage, poor yield or storage life, and lack of appeal for home use or lack of marketability. The farmer's experience enables him to rank the populations or varieties available for planting for meeting these concerns. In population genetic terms, the farmer's concerns generate a multiniche model with different populations being differentially adapted to each niche. Bellon hypothesizes that the farmer retains the variety that best meets each concern. A variety is discarded if it no longer ranks first in meeting any one of the concerns. Overall, a suite of varieties is needed to meet all of a farmer's concerns. The concerns themselves are dynamic, changing with new market structures, technology, and government policies. Bellon's model thus suggests that the focusing or narrowing of concerns at the farm level may be the trigger for loss of diversity. A challenge for this model, however, is the relative size of each niche, and the integration of concerns across the whole farm when survival — for example, during drought — becomes overriding.

Indicators of the genetic composition of landraces

The next major question is to assess the genetic diversity of populations still *in situ*. How genetically variable are the landrace populations currently growing on farm? How much do they differ in their genetic makeup from one another and from scientifically bred varieties, in terms of the particular alleles and the level of variation they contain?

Genetic diversity and divergence require assessment for two sets of attributes, analogous to the characterization and evaluation data of genetic resources. The first set is *marker diversity,* or the extent of differences between individual copies of genes. The differences should be detected as close as possible to the DNA level, for a sample of homologous sequences representative of various classes of sequences (nuclear, organellar, structural, control, spacer). This set of attributes is informative as to the ancestry or breeding history of the populations. They are indicators of the recency of bottlenecks in population size, the prevalence of outcrossing, the ease with which genes are recombined, and the level of gene flow between populations. The second set is *variation in adaptation*. This set comprises indicators of the degree to which populations are adapted to their environment and of their potential for continued performance or donors of characters in plant breeding. Both biotic and abiotic aspects of the environment are involved.

Marker diversity

Hamrick and Godt (1997) have recently summarized the published estimates of genetic diversity based upon the allozyme data for crop species. Typically in such studies, the summary measures are observed heterozygosity, expected heterozygosity (gene diversity), and number of alleles detected per locus (allelic richness). While most similar studies are based on gene bank material, breeders' collections, or cultivars, this study is particularly useful for indicating broad trends. Crop species on average have *more* genetic diversity than wild plant species at the species level, although they generally are *less* diverse than their close wild relatives. Populations of crop species are more genetically divergent among themselves than are those of wild species both in the alleles they contain and in their differences in levels of diversity between them. In broad terms, this reflects the effect of breeding system, range expansion, and diversifying selection through human agency.

Examples of studies of marker diversity (allozymes, RFLPs, RAPDs) in landraces are those in barley (Brown and Munday 1982; Bekele 1983a, b; Demissie and Bjornstad 1997); maize (Doebley et al. 1985; Geric et al. 1989); and cotton (Brubaker and Wendel 1994). Bekele (1983a) estimated allozyme diversity in 158 landrace populations of barley in Ethiopia. About 20 individuals per population from 72 areas distributed among 19 regions in Ethiopia were assayed for isozyme variation at five loci. Diversity is measured as the probability that two seeds drawn from within a population, or from two different populations, etc., will differ at a locus. The diversity had a well-developed hierarchy with average gene diversity within localities of 0.163; between localities within areas of 0.236; between areas within regions of 0.304; and, at the broadest level, between regions of 0.363. Assuming equilibrium under the island model of migration, these estimates of diversity translate to migration rates of 0.6, 0.8, and 1.3 migrants per population per generation, respectively.

Far fewer studies are available on the multilocus structure of landrace populations, that is, the extent to which genetic variants at one locus are correlated in occurrence with variants at another. Such structure arises from selection, genetic drift, or fragmentation of the population, and is retained through selection, isolation and the lack of migration, and restrictions on outcrossing and genetic recombination. Bekele (1983b) computed Brown et al.'s (1980) standardized variance measure to assess multilocus association within the regions of Ethiopia sampled. The average of the 17 median values indicated a 70% inflation of variance due to correlation of alleles at the different loci within regions. This value is comparable to that for natural populations of *Hordeum spontaneum* (80%). Part of the association among loci in landraces would be due to differentiation among the populations sampled within each region. A major cause, however, would be the "metapopulation structure" of landraces over a whole region, in which sporadic replanting (colonization), introduction from elsewhere, and migration (gene flow) oppose local extinction and divergence in individual fields. The mating system of predominant self-pollination greatly slows the decay of the resulting disequilibrium, as well as assists in the retention of any adaptive combinations of alleles at loci governing adaptive traits.

Despite the expense and effort required, estimates of marker diversity are instructive as to the "coancestry" of homologous genes in individuals and populations of landraces, and the evolutionary forces that affect the whole genome. However, not all populations of all landraces conserved *in situ* can be subject to genetic analysis. The challenge is to develop a structured representative sample of such studies, from which general extrapolation to other similar populations will be reasonably sound. The key parameters in developing such a sample would be those that Hamrick and Godt (1997) have shown as useful to structure genetic data, namely breeding system, life history, taxonomy, range, isolation, and dispersal.

Variation in adaptation

Much evidence and experience attests that landraces are adapted to their local environments (Frankel et al. 1995). If they come from marginal environments, they are known to match or better the performance of imported advanced cultivars in those marginal environments (Weltzien and Fischbeck 1990). Many studies have readily detected broad-scale geographic differences between landraces from different regions within a country [e.g., yield and seed-size lentils in Ethiopia (Bejiga et al. 1996); stress tolerance and stem solidity in durum wheat in Turkey (Damania et al. 1997)]. Weltzien (1989) analyzed the geographic patterns in barley landraces from Syria and Jordan for morphological and developmental traits. Nine groups of landraces were defined based on similarity of traits. Each group showed a close association to specific geographic or environmental factors. These results emphasize the importance of recording the locations of origin of samples and the reality of groups based on such data (see Ceccarelli and Grando, this volume).

Assessment of landrace populations for comparative yield and for components of yield is important for both the immediate local use of the material in participatory reselection and breeding programs, and the wider international valuation and use of the germplasm. For example, Moghaddam et al. (1997) analyzed the genetic variation for yield, its components, and other developmental traits in lines extracted from seven landraces of bread wheat from Iran. They found most of these characters had high levels of genetic variance. They concluded the landraces could readily be improved by identifying and intercrossing the promising genotypes.

Resistance to diseases and pests are characters of widespread use, and the number of studies of resistance in landraces to exotic or to local strains of pests is growing. Here the major research issues are the scale of pattern of variation, the relative importance of major, race-specific resistance, and the relation between reaction to exotic vs. endemic pathotypes. Pronounced patterns on a macrogeographic scale are likely, with resistance common in the same areas and lacking in others. Such patterns occur because of the conjunction of genetic diversity in both the host and the pathogen species, and an environment favorable for both. A striking example is that for resistances to rust and late leaf spot in peanuts, caused by *Puccinia arachidis* and *Phaeoisariopsis personata*, respectively. Subrahmanyam et al. (1989) screened the *Arachis hypogea* germplasm accessions conserved at ICRISAT. Some 75% of the resistant accessions originated in Peru, particularly the Tarapoto region.

Ethiopian landraces of barley have been tested for resistance against two major pathogens. For *Puccinia hordei*, Alemayehu and Parlevliet (1996) found a near absence of race-specific, major resistance and a high frequency of moderate levels of partial resistance. This showed itself as pronounced variation between and within landraces in latent period, a multigenic character.

The picture for this pathogen contrasts with that for *Erysiphe graminis hordei*, the causal agent of powdery mildew in barley. Negassa (1985) examined 421 landrace samples from 12 provinces of Ethiopia for their infection type response to seven stock cultures of powdery mildew. Resistance was prevalent: only 9% of samples were fully susceptible to all cultures and nearly 30% were resistant to all seven. About 70% had a single gene for resistance and a further 20% had two genes. The more surprising findings of this study were (1) the high frequency of populations with just a single resistance gene, implying that the pyramiding of many resistance genes was "of limited importance ... in subsistence agriculture" in this pathosystem; (2) that typically each accession was uniform in mildew reaction, implying that resistance polymorphism is an infrequent strategy; and (3) that almost all resistance genes confer incomplete resistance rather than immunity.

On the other hand, Jones and Davies (1985) tested the response of 39 old European barley varieties to powdery mildew. They were found to lack major genes for resistance (no hypersensitive seedling response). When tested for adult plant resistance in field nurseries over 3 years, the mean percent leaf damage ranged from 11 to 50%, which they suggested indicated a useful source of non-hypersensitive resistance. If, however, such resistance is multigenic, it would be difficult to breed into other cultivars.

Changes in time in population genetic structure

One advantage to conserving *in situ* that many advocate is that it provides for dynamic conservation in relation to environmental changes, pests, and diseases (Maxted et al. 1997). *In situ* conservation is a dynamic process. While the particular attributes, characters, or adaptations of a population may persist over generations, the underlying genotypes will change. New alleles or combinations are expected to arise and increase in frequency at the expense of other alleles that may well disappear. Strictly speaking, *in situ* strategies fail to preserve all the extant biodiversity at the gene level. As better alleles or combinations arise and enjoy selective advantage, others thereby will be less fit and decline. This is the cost of evolutionary substitution and the price paid for allowing evolution to continue. The likely flux of genetic variants in *in situ* strategies is of concern to some: Holden et al. (1993) argue that "museum farms" may not only fail to preserve the natural diversity that has evolved in the past, but the genetic changes in them may be unrelated to the needs of posterity.

Evidence of the nature, pace, and causation of genetic change during on-farm conservation is crucial to an understanding of on-farm conservation, and is virtually nonexistent. How rapidly do allele frequencies change, are alleles and genotypes lost, or do whole populations go extinct, locally and absolutely? What are the roles of stochastic events as opposed to systematic forces in causing such changes? (The stochastic events include bottlenecks in population size, sporadic migration, variation in mating system. The systematic forces include farmer selection, both deliberate and inadvertent, mixing, and hybridization.) What is the impact of fragmentation and decreasing area on the genetic structure of populations? What are the dynamics of seed (gene) flow between populations (see Louette, Chapter 5, this volume)?

Recognizing that fragmentation and declining area are the major trends in landrace plantings, Qualset et al. (1997) suggest that the theory of island biogeography be invoked to determine the key variables that determine the dynamics of diversity. These are patch size, frequency of migration (seed exchange between farms locally or from outside sources), and the expected positive relationships between patch (island) size or isolation and diversity.

Whereas questions of causation are perennially difficult, new technologies open up new approaches. Clegg (1997) has recently discussed the struggle to measure selection acting upon plant genetic diversity. He notes:

> The fundamental research program of population genetics has been to seek a quantitative assessment of the role of the various forces of evolution in shaping patterns of genetic variation…. New insights into the relative importance of selection and random genetic drift can now be obtained from samples of DNA sequences of genes drawn from within species. The elaboration of coalescence theory together with data on gene genealogies [from DNA sequences within and between species] permits an integration over long periods of evolutionary time [… and thus …] the detection of small selection intensities" (Clegg 1997:1).

In the future, these approaches are likely to be applicable to tracing the history and relationships of landrace populations.

Coevolutionary changes

Perhaps the major impetus for *in situ* conservation of the biodiversity of use to agriculture is the suggestion that such strategies provide the opportunity for continuing coevolution. It is argued that continuing pathogen evolution will render obsolete the samples of resistance genes "frozen" *ex situ* in gene banks. This led to the claim that such diversity would be better "conserved" *in situ*, where new resistance might evolve to match any change in virulence structure of the pathogen. What is usually overlooked is the reciprocal argument that the presence of resistance genes in landraces "unfrozen" on-farm will inevitably evoke changes in the pathogen population that could equally render the resistances obsolete.

Holden et al. (1993:90) have questioned whether on-farm conservation can evolve "novel" resistance genes, because of the evolutionary resilience in growing traditional varieties in traditional ways. Genetic heterogeneity for resistance genes is the rule. As Holden et al. note, however, "most disease and most pathogen strains are to be found in most years, but at a low level, and therefore applying low selection pressure to the resistance alleles." They contend, therefore, that "it is difficult to see how the preservation of landraces and old varieties in archaic but stable systems, can give rise to the evolution of novel resistance genes." Qualset et al. (1997) note a further point arising from the fragmentation of landrace planting. Crucial in the coevolutionary dynamic is whether the islands of landraces amid a sea of the same species act as an alternate host with a particular resistance structure or in rotation with bred cultivars of the same species in the same fields. In these host–pathogen interactions, the dispersal dynamics and survival structures of each pathogen species are critical variables. Dispersal, survival, and the pattern of host heterogeneity have a great effect on the anticipated coevolution because the pathogen population would be subject to an additional element of diversifying selection on the alternative populations of host. Clearly, the nature and pace of change of resistance structures in landrace populations conserved on farm are key topics about which there is much speculation and some dogma, but very little hard evidence.

Composite crosses

From the above discussion it is evident there is much to learn about the temporal dynamics of genetic diversity during on-farm conservation. For such research, a paradigm would be helpful. Population genetic research on the composite crosses, notably in barley, offers such a paradigm for research into the population genetics of *in situ* conservation (Suneson 1956). These are populations synthesized from a diversity of sources and then planted over many generations at one or more specific sites. Research on the composites at least shows the kinds of inference that attend a periodic sampling of generations and storage of samples for later comparison.

Allard (1988) summarized the results of long-term studies of changes in adaptedness in several barley composite cross populations. These studies included temporal changes in marker allele frequencies, in quantitative characters and fitness components, and fitness itself. His overarching generalization was that superior reproductive capacity (in terms of the number of seeds per plant) was the one quantitative character consistently associated with the increasingly prevailing allele at marker loci.

The composite cross paradigm departs, however, in some key respects from landrace populations conserved on-farm. Their origins are strikingly different, with the composites founded as a hybrid swarm of many genotypes with widely dissimilar origins. Growing such a swarm at one site inevitably leads to major changes in allele frequencies and a dramatic reduction in quantitative genetic variance. For example, Jana and Khangura (1986) report that a bulk population grown at four different sites showed loss of diversity in all populations for morphological and agronomic characters in contrast to the retention of diversity at eight isozyme loci. Allard (1988) found that while most alleles are retained in the barley composites, a few alleles increased in frequency while the remainder tended to extreme rarity. Such rarity may cause problems because it will require very large samples for detection.

Fitness also differs between cereal composites and landraces, because fecundity is simpler in the composites, whereas in landraces seed selection by farmers for quality, flavor, size, appearance, market appeal, etc. comes into play. Of course natural selection in composites can be supplemented with mass screenings for traits like seed size or cullings of heavily diseased or tall plants as parents for the next generation. Le Boulc'h et al. (1994) have drawn attention to the need for countermeasures to stop the loss of dwarfing genes from their wheat composites. But such simple steps of artificial selection hardly match the complexity of culturally based farmer selection and marker appraisal. Composites, in short, aim to give scope for recombination in the context of mass selection ("evolutionary plant breeding"), while landraces aim to produce a consumable or marketable product while conserving variation ("evolutionary sustainable production"). Both are compromises, but of two sets of different functions.

The study of the evolution of disease resistance in composite crosses is of particular interest in guiding research in on-farm conservation. Allard (1990) summarized studies of the *Hordeum vulgare–Rhynchosporium secalis* pathosystem for barley composites, emphasizing the interactive and self-regulating adjustments that occur in genetically heterogeneous populations. The pathotype structure of this pathogen is complex, comprising a wide range of abilities to damage the host. In response, the resistance allele structure in the host is also complex, with alleles differing widely in the protection they afford. Many of the resistance alleles had net detrimental effects on yield and reproductive fitness. Yet resistance alleles that protected against the most damaging pathotypes increased sharply in frequency in Composite

Cross (CC) II. These data are evidence that composites propagated under cultivation can lead to increases in the frequency of desirable alleles.

De Smet et al. (1985) examined barley composites for resistance to another major foliar disease, powdery mildew, specifically to test whether resistance is conserved. Three populations were grown for several decades in either disease-free (Montana) or disease-prone (California) environments. Four isolates that recognized the specific resistances in the founding parents were used to test seedling resistance. Overall, resistance was conserved more consistently in the California series than in the Montana site, but without the expected increase in frequency of resistance. This result is similar to that for scald resistance in CC II (Webster et al. 1986). Selection favored alleles for resistance in seasons when scald disease was prevalent, but such alleles were associated with detrimental effects on reproductive capacity in seasons that were unfavorable to scald.

In barley CC V and CC XXI, however, the same resistances showed much less change, presumably because of genotypic associations and whole-genome effects that are common in predominantly self-fertilizing populations (Burdon 1987). This raises the crucial point in researching the temporal dynamics of genetic variation in populations conserved *in situ*. A knowledge of the mating system and its variation in time is fundamental to an understanding of the system. One example is the study of Kahler et al. (1975), who measured outcrossing rates in three generations of barley CC V. They found that the rate had doubled between generations 8 and 28, indicating an evolution toward increased recombinational potential. Landrace populations are unlikely to show a steady secular trend like that in CC V, because, as noted above, they are not in the early stages of a synthesis from diverse sources. However, such populations are likely to show temporal variation in outcrossing rates with substantial effects on their genetic structure.

Pronounced population divergence was a feature of specific resistance alleles and adult plant resistance to powdery mildew in a series of wheat composites (Le Boulc'h et al. 1994). Clear relationships between virulence frequencies and resistance structure were lacking. However, multi-resistant recombinant genotypes appeared and the overall level of resistance increased, which augurs well for the rationale of *in situ* conservation.

Indicators of genetic structure

Four of the advantages of *in situ* conservation (numbered 3 to 7 in the second section above) specifically relate to the genetic structure of landraces. The following lists a series of indicators for investigating each of these advantages. The indicators range in technique from the molecular genetic to the anthropological. Many of them cannot be implemented on a broad scale in every conserved population. Yet a balanced approach to research on a representative sample of crops and farming systems is needed. If possible, the research should also consider the interaction between indicators and the various kinds of data.

Indicators for investigating population genetic structure of landraces

Allelic richness and multilocus genotypic diversity

- Population number and size or area of planting
- Mating system, degree of outcrossing
- Variation in human use of the produce (flavor, multipurpose varieties, etc.
- Number of distinct morphological phenotypes (subspecies, races, varieties)
- Morphological major gene polymorphisms (color, pubescence, etc.)
- Marker diversity (isozymes, RAPD, DNA fingerprints, DNA sequences, etc.)

Special adaptations to the local environment

- Habitat diversity
- Disease and pest occurrence or damage
- Phenological variation (maturity diversity)
- Targets or purposes of farmer selection
- Stress tolerance experiments (salinity, aridity)
- Response shown by selecting outstanding sub-lines or components
- Pest and pathogen resistance genes

Scale of localized diversity

- Topographic variation in the region
- Geographic cultural diversity, trading patterns, language groups, etc.
- Seed supply systems
- Transplantation experiments — field performance measurements
- Partition of marker diversity between different geographic scales
- Gene genealogies for tracing relationships between populations

Temporal changes in genetic composition

- Local history of varietal use, farmer selection, and perceived changes
- Extinction–recolonization cycles in the rotation of landraces in the landscape
- Comparison of stored or historic samples with current populations
- Changes in pathogen incidence, pathotype, and resistance structure
- Allele and genotype frequency changes in time

Operation of crop evolutionary processes

- Absence of factors leading to further fragmentation or loss of landraces
- Response to variation in agronomic practices

- Difference in genetic structure before and after "seed" selection by farmers
- Response to planting in disease nurseries
- Migration measured by genetic markers, or data on seed movement
- Variation in mating systems

Management of on-farm populations

A major issue facing the development of on-farm conservation is formulating the rationale for management of such populations. On this rationale will depend the extent and nature of any alteration in the planting and harvesting cycles that farmers and cooperating agencies might make. The conservation of populations has any of several possible aims:

1. Conserving the maximum number of multilocus genotypes and maximum allelic richness;
2. Safeguarding the evolutionary processes that generate new multilocus genotypes; and
3. Improving the population performance and increasing the productivity in a defined range of local environments.

These objectives are not necessarily exclusive of one another; neither are they identical, yet they are potentially conflicting goals. The first aim of conserving maximum diversity is best served by growing in a benign environment with relaxed selection. The second implies discerning and maintaining the current modes and intensity of evolutionary forces (selection, population sizes, isolation, gene flow, mating system, and recombination). The third implies seeking and implementing the appropriate plant breeding methods and selection regimes for landrace improvement in participatory breeding programs.

As far as genetic management for *in situ* conservation is concerned, the question is whether to prefer options that encourage genetic change in *in situ* populations, or options that allow it to take its course, or those that slow it down (Frankel et al. 1995). The principal cause of change can be grouped under three headings or axes, namely, the selection regime, the breeding system regime, and the population structure. The selection regime requires answers to questions such as whether disease levels or weed competition should be enhanced or reduced and whether soil infertilities should be remedied or infertile sites chosen. Recombination and the breeding system are perhaps less amenable to obvious manipulation, although Worede (1997) has noted that farmers have encouraged introgression from nearby stands of wild relatives of crops. However, population structure, which is the third axis, is controllable because it varies with population size and migration rates between populations. Frankel et al. (1995:175) assert that

> Each of the three axes needs to be assessed and the
> tempo and mode of genetic change optimized. Overall,
> three criteria ... should be met. These are (i) population
> survival; (ii) maintenance of evolutionary potential in
> the form of genetic diversity; and (iii) development of
> new genotypes.

Comparable dilemmas arise equally in the sociological aspects of on-farm conservation. Qualset et al. (1997) stress the need to conserve the agricultural system as a whole. The literal preservation of traditional agro-ecosystems in the face of modernization is not possible; indeed, such systems have always been dynamic. The challenge is to integrate the conservation of plant genetic resources with agricultural development, and in particular to conserve as much diversity as possible and the processes that give birth to it.

Sampling strategies

Sampling issues enter the conduct of on-farm conservation in several ways. Assuming that species, region, and cropping system are decided, the major questions are:

1. the number and spatial arrangement of populations within the system;
2. the population size for each generation and the number of parents contributing seed to the next generation; and
3. the size and frequency of samples for research, storage, and *ex situ* conservation, as complementary to *in situ* conservation programs.

Treatments of the sampling questions include those of Brown and Marshall (1995) for samples for *ex situ* conservation, and Brown and Weir (1983) for samples to estimate population genetic parameters. Brown and Marshall's (1995) guidelines for *ex situ* samples were to start from a minimum of about 50 individuals per population and, if appropriate, 50 populations per ecogeographic area. We then discussed how to alter these guidelines to take account of biological differences among species, specific targets of a mission, prior knowledge of levels, and patterns of genetic variation or practical requirements. The basic concept behind such a strategy is that population divergence is the key to the sampling and to the conservation value of the material. Excessive effort at any one site will seriously reduce the efficiency of the mission. A high total number of samples ensures that the variation shared throughout the region — the "rare widespread alleles" — will be captured anyway, regardless of deployment strategy (the number of sites and the number sampled at each site). If the total collection came from a single site, the diversity localized at all other sites will be lost. We contend that the divergence between sites or between populations is *the* fundamental target determining conservation strategy, even if it appears to amount to a small fraction of the total genetic variance.

Recently, Lawrence and Marshall (1997) discussed sampling sizes (of populations and individuals) for *in situ* programs. In general their treatments play down the level and significance of genetic divergence between populations and subpopulations, which leads them to reach some rather startling and potentially misleading conclusions. Thus, they question the case for conserving more than one subpopulation, which is based on geographical structure (from local selection and drift). They argue first that most of the variation of cross-pollinating species "occurs within rather than between their constituent subpopulations" (1997:108). Furthermore, they contend that conserving the variation of one population goes a long way toward conserving the variation of the species. Second, they appeal to theory showing that migration of one or two seeds per generation between subpopulations is sufficient to prevent fixation.

These two arguments are not sustainable. First, as Hamrick and Godt (1997) have shown, the populations of crop species are on the whole more divergent among themselves than are those of plant species in general. This is divergence measured by marker-gene polymorphisms as indicators of independent ancestry. Population divergence for selected quantitative traits (which Lawrence and Marshall rank more importantly) is likely to be even greater as it would stem from combining divergent ancestry with divergent ecology. Relative divergence as a proportion is not the indicator of conservation value; rather, absolute divergence is the key. Further the measures of proportionate divergence are based on identity F-statistics, whereas measures based on allelic richness are more appropriate in conservation. The fact that there is divergence at all justifies multipopulation sampling. Only if there were no divergence would the restriction of sampling to a single population be justified.

The second argument appeals to population genetic theory of migration to make such a claim. However, this theory is based on selectively neutral polymorphism. Once selection comes into play, very high levels of migration will not wipe out divergence between subpopulations. Hence for the fraction of the genome that is under selection, we should expect divergence in the face of migration. This portion of the variation is the key in determining strategies. A further point about divergence is that populations may diverge not only for the kinds of alleles they contain but also for the level of genetic variance. This is particularly the case for inbreeders (Schoen and Brown 1991). The best way to avoid an unlucky outcome of conserving a population with a below-average amount of genetic diversity is to include several populations.

Therefore, the conclusion that "when resources are limited, it might be better to concentrate on the conservation of the genetical variation of one population, rather than to disperse effort in an inadequate attempt to conserve this variation in several" (Lawrence and Marshall 1997:108) is seriously misleading. Concentrating on one population is bound to be inadequate. In contrast, dispersing effort over several judiciously chosen sites while ensuring minimum standards at each are maintained is guaranteed to sample both inter- and intrapopulation diversity.

What size should the conserved population be? Lawrence and Marshall (1997) recommend a minimum size of 5000 individuals. They deduce this figure from earlier recommendations of Frankel and Soulé (1981) for the effective population size of 500 multiplied by 10 to account for departures of actual from effective sizes. This size is required to retain quantitative genetic variation for longer term evolution and is a handy yardstick. It indicates the number of plants that ideally should contribute seed to the next generation. On the farm, the actual size will depend on many factors other than the number required to slow genetic drift to a certain level, such as field size, isolation from contaminating pollen, competing land use, other uses of the crop, seed viability, plant habit, etc. The 5000 yardstick is useful for indicating whether a given area is sufficient. From the standpoint of samples for research or gene banking, etc., it is generous, but it will ensure that very rare alleles have a chance of persisting. It is hard to understand why Lawrence and Marshall (1997:113) should conclude that "genetic diversity is more likely to be lost *in situ* than *ex situ*" with sizes of 5000 and 172, respectively. On the contrary, it is the capacity of *in situ* populations to store large number of alleles and genotypes that is its comparative advantage.

Conclusions

J. B. S. Haldane, one of the founders of population genetics, was responsible for two concepts that seem particularly relevant to on-farm conservation, namely, what was later called "genetic load" (Haldane 1937) and the "cost of evolution" (Haldane 1957). Conserving variation on the farm will entail some sort of cost, even when, as Bellon (1996) suggests, a multiniche model of diverse uses for the several populations applies. Further, if we plan for these populations to evolve new characters, then selection that renders the current, more frequent alternatives in the population less desirable will have to operate. Thus, for example, the evolution of resistance requires the presence of pathogen in abundance and the host population will likely suffer.

Diversity conserved on-farm is subject to a range of forces and is likely to be in a dynamic state. As yet, the data are far too limited to assess the various factors — human, biological, edaphic, or climatic — to determine the requirements for optimal outcomes. The challenge is to plan for assessment of these factors in relation to changes in genetic structure over time. Population biology research for *in situ* conservation thus needs to be both descriptive and hypothesis testing in order to guide technical improvement and management of landrace populations.

References

Alemayehu, F. and J.E. Parlevliet. 1996. Variation for resistance to *Puccinia hordei* in Ethiopian barley landraces, *Euphytica* 90:365–370.

Allard, R.W. 1988. Genetic changes associated with the evolution of adaptedness in cultivated plants and their wild progenitors, *Journal of Heredity* 79:225–239.

Allard R.W. 1990. The genetics of host-pathogen coevolution: implications for genetic resource conservation, *Journal of Heredity* 81:1–6.

Bejiga, G., S. Tsegaye, A. Tullu, and W. Erskine. 1996. Quantitative evaluation of Ethiopian landraces of lentil (*Lens culinaris*), *Genetic Resources and Crop Evolution* 443:293–301.

Bekele, E. 1983a. Some measures of gene diversity analysis of landrace populations of Ethiopian barley, *Hereditas* 98:127–143.

Bekele, E. 1983b. The neutralist-selectionist debate and estimates of allozyme multilocus structure in conservation genetics of the primitive landraces of Ethiopian barley, *Hereditas* 99:73–88.

Bellon, M.R. 1996. The dynamics of crop infraspecific diversity — a conceptual framework at the farmer level, *Economic Botany* 50:26–39.

Breese, E.L. 1989. *Regeneration and Multiplication of Germplasm Resources in Seed gene banks: The Scientific Background.* Rome: IBPGR.

Brown, A.H.D., C.L. Brubaker, and J.P. Grace. 1997. Regeneration of germplasm samples: wild versus cultivated plant species, *Crop Science* 37:7–13.

Brown, A.H.D., O.H. Frankel, D.R. Marshall, and J.T. Williams. 1989. *The Use of Plant Genetic Resources.* Cambridge: Cambridge University Press.

Brown, A.H.D. and D.R. Marshall. 1995. A basic sampling strategy: theory and practice. In *Collecting Plant Genetic Diversity Technical Guidelines*, L. Guarino, V. Ramanatha Rao, and R. Reid (eds.). Wallingford, U.K.: CAB International.

Brown, A.H.D. and J. Munday. 1982. Population genetic structure and optimal sampling of land races of barley from Iran, *Genetica* 58:85–96.

Brown, A.H.D. and B.S. Weir. 1983. Measuring genetic variation in plant populations. In *Isozymes in Plant Genetics and Breeding*, Part A, S.D. Tanksley and T.J. Orton (eds.). Amsterdam: Elsevier.

Brubaker, C.L. and J.F. Wendel. 1994. Reevaluating the origin of domesticated cotton (*Gossypium hirsutum*; Malvaceae) using nuclear restriction fragment length polymorphisms (RFLPs), *American Journal of Botany* 81:1309–1326.

Brush, S.B. 1995. *In situ* conservation of landraces in centers of crop diversity, *Crop Science* 35:346–354.

Burdon, J.J. 1987. *Disease and Plant Population Biology.* Cambridge: Cambridge University Press.

Clegg, M.T. 1997. Plant genetic diversity and the struggle to measure selection, *Journal of Heredity* 88:1–7.

Damania, A.B., L. Pecetti, C.O. Qualset, and B.O. Humeid. 1997. Diversity and geographic distribution of stem solidness and environmental stress tolerance in a collection of durum wheat landraces from Turkey, *Genetic Resources and Crop Evolution* 44:101–108.

Demissie, A. and A. Bjornstad. 1997. Geographical, altitude and agro-ecological differentiation of isozyme and hordein genotypes of landrace barleys from Ethiopia — implications to germplasm conservation, *Genetic Resources and Crop Evolution* 44:43–55.

De Smet, G.M.W., A.L. Scharen, and E.A. Hockett. 1985. Conservation of powdery mildew resistance genes in three composite cross populations of barley, *Euphytica* 34:265–272.

Doebley, J.F., M.M. Goodman, and C.F. Stuber. 1985. Isozyme variation in the races of maize from Mexico, *American Journal of Botany* 72:629–639.

Frankel, O.H., A.H.D. Brown, and J.J. Burdon. 1995. *The Conservation of Plant Biodiversity.* Cambridge: Cambridge University Press.

Frankel, O.H. and M.E. Soulé. 1981. *Conservation and Evolution.* Cambridge: Cambridge University Press.

Geric, I., M. Zlokolica, and M. Geric. 1989. *Races and Populations of Maize in Yugoslavia. Isozyme Variation and Genetic Diversity. Systematic and Evolutionary Studies of Crop Genepools.* Rome: IBPGR.

Haldane, J.B.S. 1937. The effect of variation on fitness, *American Nature* 71:337–349.

Haldane, J.B.S. 1957. The cost of natural selection, *Journal of Genetics* 55:511–524.

Hamrick, J.L. and M.J.W. Godt. 1997. Allozyme diversity in cultivated crops, *Crop Science* 37:26–30.

Holden, J.H.W., W.J. Peacock, and J.T. Williams. 1993. *Genes, Crops and the Environment.* Cambridge: Cambridge University Press.

Jana, S. and B.S. Khangura. 1986. Conservation of diversity in bulk populations of barley (*Hordeum vulgare* L.), *Euphytica* 35:761–776.

Jones, I.T. and I.J.E.R. Davies. 1985. Partial resistance to *Erysiphe graminis hordei* in old European barley varieties, *Euphytica* 34:499–507.

Kahler, A.L., M.T. Clegg, and R.W. Allard. 1975. Evolutionary changes in the mating system of an experimental population of barley (*Hordeum vulgare* L.). *Proceedings of the National Academy of Science, USA* 72:943–946.

Lawrence, M.J. and D.F. Marshall. 1997. Plant population genetics. In *Plant Genetic Conservation: The* in situ *Approach*, N. Maxted, B.V. Ford-Lloyd, and J.G. Hawkes (eds.). London: Chapman & Hall.

Le Boulc'h, V., J.L. David, P. Brabant, and C. De Vallavieille-Pope. 1994. Dynamic conservation of variability: responses of wheat populations to different selective forces including powdery mildew, *Genetics Selection Evolution* 26: 221s–240s.

Marshall, D.R. 1989. Crop genetic resources: current and emerging issues. In *Plant Population Genetics, Breeding and Genetic Resources*, A.H.D. Brown, M.T. Clegg, A.L. Kahler, and B.S. Weir. Sunderland, MA: Sinauer and Associates, Inc.

Maxted, N., B.V. Ford-Lloyd, and J.G. Hawkes. 1997. Complementary conservation strategies. In *Plant Genetic Conservation: The* in situ *Approach*, N. Maxted, B.V. Ford-Lloyd, and J.G. Hawkes (eds.). London: Chapman & Hall.

Moghaddam, M., B. Ehdaie, and J.G. Waines. 1997. Genetic variation and interrelationships of agronomic characters in landraces of bread wheat from southeastern Iran, *Euphytica* 95: 361–369.

Negassa, M. 1985. Geographic distribution and genotype diversity of resistance to powdery mildew of barley in Ethiopia, *Hereditas* 102:113–121.

Qualset, C.O., A.B. Damania, A.C.A. Zanatta, and S.B. Brush. 1997. Locally based crop plant conservation. In *Plant Genetic Conservation: The* in situ *Approach*, N. Maxted, B.V. Ford-Lloyd, and J.G. Hawkes (eds.). London: Chapman & Hall.

Schoen, D.J. and A.H.D. Brown. 1991. Intraspecific variation in population gene diversity and effective population size correlates with the mating system in plants, *Proceedings of the National Academy of Science, USA* 88:4494–4497.

Subrahmanyam, P., V.R. Rao, D. McDonald, J.P. Moss, and R.W. Gibbons. 1989. Origin of resistances to rust and late leaf spot in Peanut (*Arachis hypogea*, Fabaceae), *Economic Botany* 43:444–455.

Suneson, C.A. 1956. An evolutionary plant breeding method, *Agronomy Journal* 48:188–191.

United Nations. 1992. *Convention on Biological Diversity. Rio de Janiero, Brazil: United Nations Conference on the Environment and Development.*

Webster, R.K., M.A. Saghai-Maroof, and R.W. Allard. 1986. Evolutionary response of barley composite cross II to *Rhynchosporium secalis* analyzed by pathogenic complexity and gene-by-race relationships, *Phytopathology* 76:661–668.

Weltzien, E. 1989. Differentiation among barley landrace populations from the Near East, *Euphytica* 43:29–39.

Weltzien, E. and G. Fischbeck. 1990. Performance and variability of local barley landraces in Near-East environments, *Plant Breeding* 104:58–67.

Worede, M. 1997. Ethiopian *in situ* conservation. In *Plant Genetic Conservation: The in situ Approach.* N. Maxted, B.V. Ford-Lloyd, and J.G. Hawkes (eds.). London: Chapman & Hall.

Zeven, A.C. 1996. Results of activities to maintain landraces and other material in some European countries *in situ* before 1945 and what we may learn from them, *Genetic Resources and Crop Evolution* 43:337–341.

Section III

Case studies

chapter three

Barley landraces from the Fertile Crescent: a lesson for plant breeders

Salvatore Ceccarelli and Stefania Grando

Introduction

The domestication of wheat and barley took place prior to 7000 B.C. in the region of the Near East known as the "Fertile Crescent." The Fertile Crescent includes parts of Jordan, Lebanon, Palestine, Syria, southeastern Turkey, Iraq, and western Iran (Figure 3.1). Evidence suggests that the most important of the early cereals was barley, and the archaeobotanical material from the region clearly shows that the first barleys were two-rowed (Harlan and Zohary 1966). The wild progenitor of cultivated barley, *Hordeum vulgare* ssp. *spontaneum*, is still widely distributed along the Fertile Crescent where, particularly in the driest areas, it can be easily identified from a distance because of its height. It is likely that *Hordeum spontaneum* contributes to the evolutionary processes of barley landraces through a continuous introgression of genes.

Today barley is still one of the most important cereal crops in the Fertile Crescent, spanning an area of approximately 5 million hectares. Barley is a typical crop in marginal, low-input, drought stressed environments (Ceccarelli 1984). Barley seed and straw are the most important source of feed for small ruminants, primarily sheep, and therefore palatability of straw in particular, but also of grain, is an important attribute to most farmers. Conventional breeding and high yielding varieties (HYVs) have had virtually no success in this region, which has had a positive effect on preserving biodiversity. In these environments, all cultivated barleys are landraces (Weltzien 1988) that have evolved directly from the wild progenitor. They have adapted to hostile environments and are popular among farmers for their high feed quality as both grain and straw.

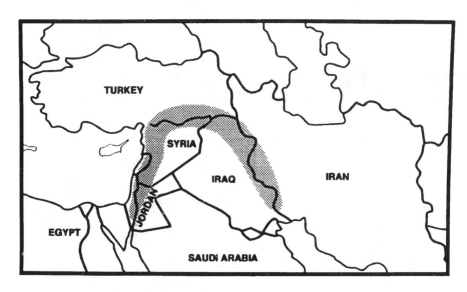

Figure 3.1 The "Fertile Crescent," where crops such as barley, wheat, lentil, stone fruits, and olives were domesticated. (Modified from Harlan and Zohary 1966.)

In Syria farmers identify two major groups of landraces, largely on the basis of seed color, namely Arabi Abiad (white seed) and Arabi Aswad (black seed). Arabi Abiad is common in environments receiving between 250 and 400 mm annual rainfall; Arabi Aswad is cultivated in harsher environments with less than 250 mm annual rainfall. Although Vavilov had collected these two barley landraces by the beginning of the century, little is known about them. A few accessions have been included in the world collection, but as with many other crops no use has been made of these valuable genetic resources.

In the early 1980s, it was postulated that because barley landraces have been grown continuously since domestication without inputs in unfavorable and stress environments, their evaluation could teach a barley breeder a few lessons about adaptation to low-input, stress environments. It was also postulated that these lessons could prove useful to other breeders in countries where barley landraces are still predominant, as well as to breeders of crops mostly cultivated in stress environments (Ceccarelli 1984).

The objective of this chapter is to illustrate the use of landraces in the barley breeding program at the International Center for Agricultural Research in the Dry Areas (ICARDA), as an example of the contribution that landraces can make to increasing agricultural production, particularly for the rural poor in marginal environments. Implicitly, these arguments suggest that securing the continuity of the evolutionary processes within landrace polulations is of vital importance for future generations.

Collection and preliminary evaluation

In 1981, E. Weltzien made an extensive collection of barley in Syria and Jordan (Figure 3.2) from the fields of 70 farmers (60 in Syria and 10 in Jordan), who had been using their own seed for generations. One hundred spikes were collected at random from each farmer's field (Weltzien 1988). The spikes were kept separate, contrary to most conventional collection methods. This was a key factor in the subsequent utilization of the collection.

When the collected seed was multiplied off-season (planting in summer) as individual rows, each planted with the seed of one spike (head-rows), two main characteristics were noted. First, a high degree of seed dormancy was observed, with the material collected in southern Jordan showing a higher percentage of germination. Second, few of the rows were able to head and produce seed, with differences in the material collected at the same sites (Weltzien 1982). Additional information on the structure of the variation between and within collection sites was obtained when the material was evaluated under field conditions as individual rows (Weltzien 1988, 1989) or as plots (Weltzien and Fischbeck 1990). Significant genetic variation was found for seed color, growth habit, awn barbing, days to heading, culm length, leaf width, awn length, early growth vigor, lodging score, and powdery mildew resistance.

We recognize now that these were the first lessons the landraces were teaching, both by indicating traits of adaptive significance (such as vernalization requirement and seed dormancy) and by expressing the variability harbored within these populations. Three important findings, which were later confirmed, emerged from this preliminary evaluation. First, the genetic variability within the landraces was expressed in stress sites, where the heritability was even higher than in a non-stress site. Second, in a stress site the majority of landraces outyielded the check (improved) cultivars. Lastly, in the non-stress site the checks outyielded the landraces, though not always significantly.

From preliminary evaluation to breeding

Prior to 1984, the barley breeding program at ICARDA did not utilize landraces in a systematic fashion (Ceccarelli 1984), although the preliminary data were extremely promising, as indicated above. The procedure for utilizing the material of the barley landrace collection was first to assess the amount of genetic variation for agronomic and morphological characteristics, and then to determine the extent to which genetic diversity within the landraces was useful for breeding purposes. We focused attention primarily on Arabi Abiad and Arabi Aswad, the two barley landraces most widely grown in Syria.

In 1984, the barley breeding program began testing all of the breeding materials under typical growing conditions for barley in Northern Syria: strictly rainfed, predominantly in areas with low and erratic rainfall, and with little, if any, use of fertilizers, pesticides, or herbicides. The strategy was based on the assumption — later proven to be correct — that useful genetic variation for stress conditions could only be detected by testing breeding material under farmers growing conditions. To achieve this, we rented a farmer's field in an area that the breeding program had not previously used, referred to as Bouider, and we expanded the work already underway at Breda. Together with the experiment station at ICARDA headquarters in Tel Hadya, the experiment sites represent three distinct agricultural systems. Tel Hadya is a favorable high-input environment which lends itself to a wide choice of different crops. Bouider represents the opposite extreme: a typical low-input, high risk environment where barley is the only rainfed field crop. Breda is intermediate between the two, located at the beginning of the area where Arabi Aswad becomes the dominant landrace. The three sites are geographically close, located at 35 (Tel Hadya), 60 (Breda), and 80 km (Bouider) southeast of Aleppo, which provides an enormous advantage in terms of field operations. Table 3.1 shows the total rainfall at the three sites since the work on landraces began. Although rainfall does not convey all the information about climate — rainfall distribution and winter temperatures also play a determinant role — it is evident that there is a consistent rainfall gradient between the three sites, which makes the area unique in providing large climatic contrasts within short distances.

Table 3.1   Total Rainfall (mm) in the Three Experimental Sites Used by the Barley Breeding Program in Northern Syria

Year	Tel Hadya	Breda	Bouider
1984–1985	372.6	276.6	—
1985–1986	316.4	218.3	203.0
1986–1987	357.9	244.6	176.2
1987–1988	504.2	414.0	385.7
1988–1989	234.4	194.8	189.0
1989–1990	233.4	183.2	148.7
1990–1991	293.5	241.3	213.4
1991–1992	352.6	263.2	249.6
1992–1993	390.1	283.0	224.2
1993–1994	373.3	291.2	245.6
1994–1995	312.9	244.2	203.1
1995–1996	404.5	359.8	316.0
Long Term	328.9	267.8	235.2

In the season 1984–1985, 420 single-head progenies (lines) were evaluated at Breda in three trials. In the first two trials, we evaluated 280 lines representing 28 collection sites with 10 lines per collection site (Ceccarelli et al. 1987). Each trial contained 140 lines (10 for each of 14 collection sites) and four checks (Arabi Abiad and Arabi Aswad, and two improved cultivars

Harmal and Rihane-03). In the third trial, we evaluated 70 lines for each of two collection sites. Because the amount of seed was still a limiting factor, the lines were planted in two-row plots in the first two trials, and in four-row plots in the third trial. The following characters were measured or scored: growth habit, early growth vigor, cold damage, plant height, days to heading, days to maturity, grain filling duration, grain yield, spike length, peduncle extrusion, 1000 kernel weight, protein content, lysine content, and seed color.

Not surprisingly, a large and significant variability was found for virtually all of the characters measured. The mean squares between collection sites were nearly always significantly larger than the error mean square (Table 3.2). Also the variation within collection sites was almost always significantly larger than the error term. The "between collection site" component was in most cases significantly larger (P < 0.01) than the "within site" component.

Table 3.2   Mean Squares between and within Collection Sites for Agronomic Characters in Two Experiments with Single-Head Lines Derived from Local Cultivars

	Experiment 1			Experiment 2		
Character	Between Sites	Within Sites	Error	Between Sites	Within Sites	Error
Growth habit	6.97**	0.36**	0.2	5.05**	0.28**	0.2
Cold damage	8.08**	0.95**	0.49	9.36**	0.45	0.38
Days to heading	7.55**	4.86**	1.83	135.92**	8.90**	2.12
Days to maturity	8.96**	2.41*	1.58	115.45**	2.87**	1.62
Grain filling	3.28**	2.66**	1.71	22.14**	6.91**	2.29
Plant height	486.59**	43.44**	22.25	724.68**	36.06**	13.35
Spike length	0.83*	0.45*	0.3	5.78**	0.93**	0.4
Peduncle extrusion	64.20**	13.24**	3.15	141.43**	9.27**	3.42
1000 KW	168.88**	18.77**	4.8	270.06**	15.35**	5.22
% protein	1.26**	0.71*	0.49	3.95**	0.66*	0.43
% lysine[a]	0.5	0.50**	0.29	0.92**	0.43**	0.25
Grain yield[b]	53.77**	4.87	4.59	67.13**	5.14**	3.25

* P < 0.05

**P < 0.01

[a] $(\times 10^{-3})$

[b] $(\times 10^{3})$

These data also quantified some of the key differences between the white-seeded and the black-seeded landraces (Table 3.3). These differences are of particular interest to plant breeders in verifying the firm belief of Syrian farmers that the black-seeded landrace is better adapted to dry areas and provides better feed for sheep than the white-seeded landrace. In this case, the use of lines with specific seed colors could become important to ensure quick adoption. Using the data of the third experiment we found that the black-seeded landrace is usually less vigorous in early growth, more cold tolerant and more productive under stress than white-seeded landrace. Arabi Aswad matures slightly earlier and has a shorter grain filling period than Arabi Abiad. Finally, plants tend to be taller, have smaller kernels, shorter coleoptile length, and shorter and fewer seminal roots. Some of these differences, such as those associated with phenology, cold tolerance, growth vigor, kernel size and plant height, are related to adaptation to dry and cold areas where the black type is predominantly cultivated. Syrian farmers often note the advantage of plant height under conditions of drought as one of the main reasons for preferring Arabi Aswad to Arabi Abiad in the drier areas. One of the primary effects of drought is a drastic reduction of plant height and a consequent reduction of straw yield. This increases the cost of harvesting, as it must be done by hand rather than by combine.

Table 3.3   Differences between the Black-Seeded (Arabi Aswad) and the White-Seeded (Arabi Abiad) Barley Landraces Commonly Grown in Syria

	Arabi Aswad (Black)		Arabi Abiad (White)	
Character	Mean ± s.e.	Range	Mean ± s.e.	Range
Growth vigor[a]	3.01 ± 0.09	4.48–1.35	3.69 ± 0.10	4.94–0.98
Cold damage[b]	2.10 ± 0.06	3.08–1.02	3.26 ± 0.08	4.68–1.58
Days to heading[c]	147.4 ± 0.22	153.0–141.5	147.7 ± 0.14	150.4–145.03
Days to maturity[c]	171.9 ± 0.30	177.6–168.8	173.8 ± 0.28	178.5–168.8
Grain fill. duration (days)	24.5 ± 0.26	30.5–19.50	26.1 ± 0.25	30.5–20.5
Plant height (cm)	52.1 ± 0.47	61.8–40.9	43.1 ± 0.46	53.4–33.4
Grain yield (kg/ha)	1769 ± 36	2480–944	1542 ± 40	2324–920
Protein content (%)	10.5 ± 0.05	11.6–9.7	10.6 ± 0.06	11.9–9.9
Lysine content (%)	0.43 ± 0.00	0.45–0.41	0.43 ± 0.00	0.46–0.40
1000 kernel weight (g)	35.7 ± 0.29	43.5–31.1	41.9 ± 0.35	47.9–34.6
Root number	57 ± 0.06	7.1–4.4	6.2 ± 0.05	7.4–5.0
Root length (mm)	55.8 ± 1.23	86.3–37.16	69.1 ± 1.11	99.3–43.3
Coleoptile length (mm)	47.5 ± 0.44	55.4–39.4	52.4 ± 0.55	61.4–41.4

[a] 1 = poor; 5 = good

[b] 1 = minimum; 5 = maximum

[c] Days from emergence

The most interesting aspect of this early work was the extraordinary amount of variability found *within* landraces as shown by the analysis of variance (Table 3.2) and the interval of variation (Table 3.3). The observation that landraces are composed of several genotypes is neither new nor original and has been reported for several crops, such as lentil (Erskine and Choudhary 1986), sorghum (Blum et al. 1991), bread and durum wheat (Porceddu and Scarascia Mugnozza 1984; Damania and Porceddu 1983; Spagnoletti-Zeuli et al. 1984; Damania et al. 1985; Lagudah et al. 1987; Blum et al. 1989; Elings and Nachit 1991), beans (Martin and Adams 1987a, 1987b), and both cultivated and wild barley (Brown 1978, 1979; Asfaw 1989).

In the case of Syrian barley landraces, the presence of such a high level of heterogeneity is not as obvious at first sight as it is, for example, in Ethiopian or Nepalese barley landraces. This hidden morphological variability might explain why Syrian farmers do not select within landraces either before or after harvesting but are able to distinguish between cultivars. Also, one could hypothesize that thousands of years of natural and human selection in a stress environment could have reduced the amount of heterogeneity through continuous selection for the most adapted genotypes. Not only does this not seem to be the case, but the variation available within the population appears to be large and of great value to a breeding program for stress environments and low-input conditions. This is most strongly indicated by the yield advantage of some of the pure lines extracted from landraces over both original landraces and some improved (modern) cultivars (Table 3.4).

Table 3.4   The Highest Yielding Pure Lines Extracted from Landraces in 1984–1985 in Breda (277 mm rainfall) Compared with the Two Commonly Grown Landraces (A. Abiad and A. Aswad) and Two Improved Cultivars (Rihane-0.3 and Harmal)

Entry	Seed Color	Plant Height (cm)	Grain Yield (kg/ha)
SLB 45-48	black	50.8	2480
SLB 39-31	white	42.4	2324
SLB 39-58	white	45.1	2287
SLB 45-83	black	55.9	2232
SLB 45-95	black	53.7	2227
SLB 45-40	black	61.5	2216
SLB 39-05	white	45.3	2189
SLB 45-04	black	55.2	2180
SLB 39-10	white	45.0	2162
SLB 45-90	black	61.8	2153
SLB 45-34	black	53.1	2146
SLB 45-76	black	53.6	2122
Checks			
A. Abiad (landrace)		45.4	1666
A. Aswad (landrace)		47.7	1547
Rihane-03 (modern)		49.4	1013
Harmal (modern)		45.9	1017
$LSD_{0.05}$		5.4	453

The data show a considerable yield advantage of the landraces over modern varieties in low rainfall conditions and with little or no use of inputs. These data have been confirmed in many comparisons between different types of germplasm in such an environment (Ceccarelli and Grando 1996) and, in part, explain the failure of introducing modern cultivars into the area. The most important information from a breeding point of view regards the amount of improvement which can be achieved by simply utilizing the variability present within landrace populations.

In addition to grain yield and other agronomic, morphological, and physiological characters, an unexpected amount of variability was found for disease resistance, particularly for yellow rust, powdery mildew, scald, and covered smut (Table 3.5) (van Leur et al. 1989). With the exception of covered smut, there was a significant variation both between and within collection sites. The response to diseases varied from absolutely or partially resistant types to highly susceptible lines. These findings challenge the common belief that landraces are disease susceptible, and therefore not worth the attention of modern plant breeders. The data (Table 3.5) indicate that although landraces appear disease susceptible because the majority of plants are susceptible, they do contain a small frequency of resistant individuals that are an important source of genes for disease resistance within an adapted genetic background.

Table 3.5   Mean Squares of Combined Analysis of Variance of Disease Readings on 140 Pure Lines Collected from 14 Collection Sites (10 lines per collection site) over 2 years

Source of Variation	df	Yellow Rust	Powdery Mildew	Scald	Covered Smut
Years	1	14781***	1215.40***	534.41***	2984.0***
Lines	139	1805***	10.92***	6.38***	47.1
Collection sites	13	7682***	63.28***	22.84**	133.7
Lines within co. sites	126	1199***	5.51**	4.68***	38.2
Lines × years	139	357**	4.1	2.48**	38.0***
Coll. sites × years	13	708***	9.40**	5.87***	108.1***
Lines w. sites × years	126	321*	3.56	2.13	30.7***
Residual	278	234	3.52	1.72	8.2

* $P < 0.05$
**$P < 0.01$
***$P < 0.001$

The presence of a high level of genetic diversity within populations — adapted to an environment where conventional breeding has failed — suggests that in addition to the need for continuous collection and both *ex situ* and *in situ* conservation, there is the almost unexplored possibility of using this large reservoir of genetic variation for plant improvement. To investigate further, we identified four strategies:

1. Develop highest yielding pure lines extracted from landraces into pure line varieties, after testing their stability in different environments (across sites and years);
2. Utilize pure lines extracted from landraces, which are superior for yield as well as for other characters including quality and resistance to insect pests and diseases, as parents in the crossing program to introduce additional desirable characters in an adapted genetic background;

3. Develop mixtures or multi-line varieties, constructed with a variable number of pure lines properly characterized for a set of agronomic characters. This permits us to exploit the buffering capacity of genetically heterogeneous populations in relation to stability and will conserve a certain amount of the evolutionary process within populations;

4. Evaluate lines with contrasting expressions of specific characters to quantify their adaptive role in stress environments, and use molecular techniques to identify, localize, and tag gene complexes or loci controlling quantitative traits (Quantitative Trait Loci or QTL) associated with adaptation.

The first three strategies aim to directly utilize the genetic variability within landraces, while the fourth aims to illustrate the usefulness of landraces as a unique source of information on mechanisms of adaptation to marginal environments, stress conditions, and low-input agriculture.

ICARDA initiated each of the four strategies within a few years of each other, with the exception of the molecular approach which began only recently. It was obvious from the beginning that these activities had two main objectives: to generate new cultivars for the dry areas of Syria; and to develop a methodology for landrace utilization which could be adopted with suitable modifications for other regions and crops where landraces are still available. To achieve the second objective, we designed the methods for exploiting the genetic variation between and within landraces with the expectation that they could be used by breeders in developing countries with limited resources. A key aspect of the methodology was to implement it with the same level of inputs used by farmers in resource limited environments. This would ensure that the products (pure lines and mixtures) would be beneficial to poor farmers and yield increases could be sustained.

Landraces as breeding material

Pure line selection: the short-term approach

Since 1985, we have systematically evaluated the collection of 7000 spikes described above, using a pure-line selection method to test between 300 and 400 lines each year under typical farmers' conditions. Farmers were invited to visit the plots and to make their own selection: their selection criteria (tall plants under drought and soft straw) were subsequently incorporated into breeders' criteria.

Twelve years after the initiation of the landrace breeding program, three quarters of the collection has been evaluated, three pure lines (two black-seeded lines, Tadmor and Zanbaka, and one white-seeded, named Arta — the only line officially released) are already growing in farmers' fields on an area of 500 to 2000 hectares each. Before 1981, Tadmor, Zanbaka, and Arta were three spikes among millions from the three collection sites, indicated in Figure 3.2 with the numbers 3 (central region), 42 (northeastern region), and 39 (southern region), respectively. Today, the progenies of those three spikes are growing in farmers' fields and outyield the local landraces by 10 to 25% without additional inputs.

Figure 3.2   Geographical distribution of the collection sites of the barley landraces in Syria and Jordan.

Figure 3.3 provides an example of the yield advantage which can be obtained in farmers fields with this strategy. Arta was compared with the local landrace (either Arabi Abiad or Arabi Aswad, depending on the location) in 69 farmers' fields in five provinces of Syria. The locations have been ranked in ascending order according to the yield of the local landrace. The superiority of Arta is larger at low yield levels than at higher yield levels: in the 23 lowest yielding locations, Arta always outyielded the local landrace — yields were similar in only one case — which suggests that Arta is especially beneficial to farmers in difficult environments. Arta was already showing its superiority when tested for the first time in the season 1984–1985 (SLB 39-58 in Table 3.4).

The evaluation of the landrace collection continues to generate new and useful lines every year. In 1994, for example, we evaluated all lines from four collection sites (Figure 3.2), one with white seed (site 24), and the other three with black seed (sites 21, 22, and 23). Even when improved lines, such as Arta and Zanbaka, are used as checks for grain yield and plant height, respectively, it is possible to find lines outyielding Arta by 36% in Breda and by 13% in Tel Hadya (Table 3.6). In terms of plant height it was possible to find lines significantly taller than Zanbaka in the three collection sites with black seed, and lines taller than Arta in the collection site with white seed.

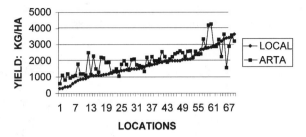

Figure 3.3 Grain yield of Arta compared with local barley in 69 farmers' fields in five provinces of Syria in 1996. Each cultivar was grown on plots of 1 ha.

Table 3.6 Variability between and within Four Collection Sites (see Figure 3.2) for Grain Yield in Two Locations, Days to Heading and Plant Height in 1994

Collection Site[a]	Breda (kg/ha)	Tel Hadya (kg/ha)	Heading	Plant Height (cm)
Site 21 (n = 86)				
means ± s.e.	1289 ± 24	2903 ± 34	111 ± 0.2	45 ± 0.4
min	891	2207	105	30
max	1837	3695	114	54
Site 22 (n = 79)				
means ± s.e.	1311 ± 21	2870 ± 26	111 ± 0.2	47 ± 0.5
min	706	2207	106	39
max	1754	3497	114	56
Site 23 (n = 70)				
means ± s.e.	1296 ± 23	2846 ± 51	110 ± 0.2	43 ± 0.6
min	832	1553	107	28
max	1837	3725	115	55
Site 24 (n = 64)				
means ± s.e.	1385 ± 25	3566 ± 54	110 ± 0.2	34 ± 0.6
min	884	1774	105	25
max	1823	4491	113	50
Checks				
Arabi Abiad	1283	3489	105	36
Arabi Aswad	1108	2799	110	44
Arta	1352	3984	106	32
Zanbaka	1110	2744	109	50

[a] The number of lines evaluated (in parentheses).

The evaluation of the landrace collection has led to two primary successes. First, we have developed three varieties that have rapidly spread from farmer to farmer. Second, over the past 10 years we have identified, within the landraces, sources of resistance to most of the major barley diseases such as powdery mildew, scald, yellow rust, covered smut, barley stripe, and root rot, which have been selected for use as parental stocks (see next section). In some cases, such as the scald resistance of Tadmor, there is strong evidence that the resistance is not based on major genes and is therefore likely to be more durable.

The evaluation of pure lines described in this section is not a separate activity, but is conducted within the context of ICARDA's barley breeding program. Therefore, it has been possible during the years to make several comparisons between the landraces of Syria and Jordan and modern cultivars. In one such study (Table 3.7), 77 lines from Syrian landraces were compared with modern cultivars using the average grain yield of two stress sites (YS) and the average grain yield of three non-stress sites (YNS). The landraces have an average yield advantage of 60% under stress while the modern cultivars have an average yield advantage of 14%. In addition to the mean performance of the two types of germplasm, the interval of variation is very informative. All 77 lines from landraces yielded something under stress, while some of the modern cultivars failed; the best modern cultivars yielded almost as much as the best landraces. Under non-stress conditions, it was interesting to find that the yield of some landraces was not significantly inferior to that of the best modern cultivars.

Table 3.7   Grain Yield (kg/ha) under Stress (YS) and Grain Yield under Non-Stress (YNS) of Barley Landraces and Modern Cultivars in Syria

Type of Germplasm	N[a]	YS[b]		YNS[c]	
		Yield	Range	Yield	Range
Modern	155	488	0–893	3901	2310–4981
Landraces[d]	77	788	486–1076	3413	2398–4610
Best check		717		4147	

[a] Number of entries;

[b] Average of two stress sites;

[c] Average of three non-stress sites;

[d] Pure lines obtained by pure line selection within landraces.

The superiority of landraces does not depend on which improved germplasm is used in the comparison, or on the specific stress environment. For four breeding cycles, each containing different breeding lines, six-row genotypes unrelated to Syrian landraces were compared with two-row genotypes, which include both modern cultivars and Syrian landraces (Table 3.8). Under stress the two-row genotypes always yielded more than the six-row genotypes with a yield advantage ranging from 15 to 38%. This advantage is largely attributed to the landraces which, under stress, have a yield advantage of 28 to 54% over modern six-row types, and one to 35% over modern two-row types. When we compared the different types of germplasm for yield potential, the landraces are always the lowest yielding type of germplasm. In the dry areas of Syria, however, the probability of yields exceeding

3 tons per ha is about six times lower than the probability of yields less than 1.5 tons per ha (Ceccarelli 1996). Therefore, the lower yield potential of the lines extracted from landraces, specifically selected for stress conditions, is not a serious problem.

Table 3.8   Yield Potential and Yield under Stress of Six- (6) and Two-Row (2) Barley Genotypes; the Two Row are Classified as Improved (I) and Landraces (L) (Number of Genotypes in Brackets)

Set[a]	Row Type	Yield Potential		Yield under Stress	
		kg/ha[b]	6R = 100	kg/ha[c]	6R = 100
1989	6 (97)	5385 ± 64	100.0	561 ± 22	100.0
	2 (203)	5135 ± 56	95.3	644 ± 13	114.8
	2L (51)	4470 ± 87	83.0	759 ± 20	135.3
	2I (126)	5396 ± 67	100.2	608 ± 16	108.4
1990	6 (120)	3975 ± 83	100.0	458 ± 15	100.0
	2 (160)	3592 ± 81	90.4	632 ± 12	138.0
	2L (86)	43170 ± 87	79.8	705 ± 12	153.9
	2I (58)	4245 ± 138	106.8	521 ± 20	113.8
1991	6 (80)	4801 ± 68	100.0	754 ± 19	100.0
	2 (120)	4808 ± 50	100.2	955 ± 12	126.7
	2L (18)	4641 ± 154	96.7	966 ± 21	128.1
	2I (102)	4837 ± 52	100.8	952 ± 13	126.3
1992	6 (22)	4504 ± 82	100.0	440 ± 38	100.0
	2 (42)	4564 ± 89	101.3	575 ± 17	130.7
	2L (11)	4376 ± 72	97.2	661 ± 19	150.2
	2I (24)	4586 ± 46	101.8	558 ± 24	126.8

[a] Each set includes breeding lines and lines from landraces evaluated for 3 years. For each example the 1989 set contains lines evaluated in 1987, 1988, and 1989 in a number of locations.

[b] Average grain yield in those year-location combinations where the grain yield of all the breeding lines was one or more standard deviations higher than the average grain yield across all the year-location combinations of that set.

[c] Average grain yield in those year-location combinations where the grain yield of all the breeding lines was one or more standard deviations lower than the average grain yield across all the year-location combinations of that set.

One of the most important messages of the data shown in Tables 3.7 and 3.8 concerns the choice of the selection environment. It is clear from the two examples that, had the selection been done only under the high yielding conditions of a typically high input experiment station, the landraces would have had a short life as breeding material. As pointed out earlier, pure lines should be only one intermediate product in the overall strategy of using landraces in a breeding program. The value of some pure lines extracted from landraces underlines the importance of *in situ* conservation programs for maintaining those processes which can continuously produce new superior genotypes within landraces. The exploitation of the variability available within landraces is a simple and efficient way to improve the productivity of crops for which landraces are still available. Similar approaches to barley selection under low-input, stressed environments are currently underway in Ethiopia (Lakew et al. 1997), Tunisia, and Iraq. New collections of barley landraces have been made recently in Nepal and Eritrea to begin landrace improvement programs. Because of its potential for increasing crop production, however, using landraces as breeding material may lead to the replacement of landraces with improved pure lines, thereby endangering the evolutionary processes on which the success of the methodology is based.

Crosses: building on adaptation

Following from the identification of agronomically superior pure lines and sources of disease resistance within landraces, we initiated the second strategy to utilize pure lines as parental material in the breeding program. An example of the value of this approach is given in Table 3.9 where 514 breeding lines unrelated to landraces (improved) and 525 pure lines extracted from landraces are compared to lines derived from three types of crosses. The data were collected in a very dry site and year (Breda received 244 mm rainfall in 1995) where we measured grain yield, total biological yield, plant height, and harvest index, and in a relatively wet site (Tel Hadya with 313 mm rainfall) where we measured yield potential. As indicated earlier, the landraces yielded on average more than the improved lines under stress and had a lower average yield potential. Under stress, landraces and improved lines had a similar biological yield, but the landraces were, surprisingly, much shorter and had a higher harvest index — two characteristics usually associated with high yielding varieties when grown under optimum conditions.

Crosses between landraces and improved germplasm generated breeding material equal to landraces in terms of grain yield and total biological yield under stress, and superior for plant height while maintaining a relatively high harvest index. Crosses between landraces and the wild progenitor of cultivated barley, *Hordeum spontaneum*, generated breeding material which is almost as good as that derived from crosses between landraces and improved germplasm. In this type of cross, the total biological yield and plant height are greater than in any other material, and both grain yield under stress and harvest index are probably underestimated because of the presence of some brittle-rachis genotypes. The last type of cross — improved × *Hordeum spontaneum* — generated the least promising type of breeding material, except perhaps for plant height.

Of the three types of crosses, crossing landraces with *H. spontaneum* has been the most promising avenue to improve plant height under drought: both plant height and straw softness are often indicated by farmers as the most desirable traits, particularly in dry areas. As mentioned earlier, a crop that remains tall even in dry years is important to farmers, because it reduces

Table 3.9 Grain Yield (kg/ha), Biological Yield (kg/ha), Plant Height (cm), and Harvest Index in Breda (1995) and Grain Yield in Tel Hadya 1995 (kg/ha) of Different Types of Breeding Material

Breeding Material	Grain Yield (BR95)[a]	Biological Yield (BR95)[a]	Grain Yield (TH95)	Plant Height (BR95)	Harvest Index (BR95)
Improved (n = 514)					
Mean	591 ± 8	1559 ± 17	4125 ± 27	23.2 ± 0.2	22.8 ± 0.3
Max	1201	4504	5812	40.3	41.3
Min	69	1559	1375	14.8	3.24
Improved ¥ Landraces (n = 214)					
Mean	775 ± 10	2678 ± 24	3883 ± 33	25.1 ± 0.3	29.1 ± 0.3
Max	1252	3658	5206	38	37.9
Min	259	1930	2630	16.9	11
Landraces (n = 525)					
Mean	752 ± 7	2549 ± 16	3657 ± 23	21.4 ± 0.1	29.8 ± 0.2
Max	1232	4027	5455	30.5	39.9
Min	320	1529	2250	13.1	16.5
Landraces ¥ *Hordeum spontaneum* (n = 133)					
Mean	724 ± 11	2829 ± 32	2797 ± 49	29.1 ± 0.4	25.9 ± 0.3
Max	1077	4007	4489	43.6	35.6
Min	369	2060	1515	20.5	11.5
Improved ¥ *Hordeum spontaneum* (n = 17)					
Mean	537 ± 37	2362 ± 111	2814 ± 118	27.1 ± 1.7	20.6 ± 1.2
Max	907	3681	3995	44.1	30
Min	306	1842	1780	19.2	11.4

[a] BR95 = Breda; TH95 = Tel Hadya 1995.

their dependence on costly hand harvesting, while soft straw is considered important in relation to palatability. Of 1532 lines tested at Breda in 1995, the mean plant height was 23.5 cm, the shortest lines were only 12.5 cm tall, and the most widely cultivated landrace (Arabi Aswad) grew to a height of roughly 25 cm (Table 3.10). Some of the lines derived from crosses with *H. spontaneum* were taller than 40 cm. They were also significantly taller than Zanbaka, the pure line selected from Arabi Aswad (described earlier), which is already grown by some farmers for its plant height.

The characteristics of height and straw texture represent a drastic departure from the typical selection criteria used in breeding high-yielding cereal crops which favors short plants with stiff straw and high harvest index. Cultivars possessing the two characteristics considered important by farmers in dry areas would be unsuitable for high-yielding environments because of their lodging susceptibility, and would not be made available to farmers in

Table 3.10   Plant Height at Breda (244 mm rainfall) in 1995 of Barley Lines Derived from Crosses with *H. spontaneum*, Compared with the Barley Landrace Most Common in Dry Areas (Arabi Aswad) and with a Cultivar Selected Specifically for Plant Height under Drought (Zanbaka)

Cross/Name	Plant Height (cm)
H. spontaneum 20-4/Arar 28//WI2291/Bgs	43.5
SLB 45-40/*H. spontaneum* 41-1	43.0
Zanbaka/*H. spontaneum* 41-2	42.5
Zanbaka/*H. spontaneum* 41-2	41.5
Moroc 9-75/Arabi Aswad//*H. spontaneum* 41-3	41.0
Arabi Aswad	24.8
Zanbaka	26.0
Mean of all breeding lines	23.5
Maximum	43.5
Minimum	12.5
$LSD_{0.05}$	5.6

a traditional breeding program — a further indication of the importance of specific adaptation.

Eventually, an interesting pattern emerged in a number of experiments: not only under drought conditions did crosses with landraces largely out-yield crosses without landraces, but crosses with specific lines from land-races, such as Tadmor (Table 3.11), were superior to all other types of crosses. This might suggest the presence of blocks of genes in chromosomal regions with low frequency of recombination conferring a specific adaptation to stress environments — a hypothesis that will be tested with the techniques of molecular genetics.

Table 3.11   Yield under Drought Stress of Crosses with Tadmor and Crosses without Tadmor

Type of Cross	Grain Yield (kg/ha) under Stress
Crosses with Tadmor	1237
Crosses without Tadmor	604

The superiority of the crosses with landraces suggests that the strategy of using adapted germplasm in a breeding program is to capitalize on their specific adaptation to drought and low-input conditions rather than to consider them as sources of new useful genes as is the case in most plant breeding programs. Therefore, in breeding for stress environments, landraces should be regarded as recipients of few useful genes to be added to their adapted genetic background, rather than as donors of traits not available in "elite germplasm." This is conceptually similar to what breeders in favorable environments do: breeders find genotypes with high yield potential and good adaptation to high-yielding conditions and continue to build on them. The strategy is strengthened by the availability of genes for disease resistance within landraces. If a line extracted from landraces is agronomically superior but susceptible to a disease, the source of resistance is first sought among lines from the same collection site to preserve as much adaptation as possible, and secondly sought among lines from neighboring collection sites. Sources of disease resistance from germplasm adapted to different environments is the last resource. For these reasons, the best germplasm pool for the Fertile Crescent is now derived from crosses involving lines extracted from landraces.

Our assessment of the value of lines extracted from landraces as parental material in a conventional crossing program is based on those lines collected in 1981 and maintained *ex situ*. Lines with higher than average breeding value (defined as the value of an individual judged by the mean value of its progenies) are presumably being continuously produced by a combination of natural and human selection and by naturally occurring intercrossing. Thus, *in situ* conservation becomes essential to ensure that the flow of superior genetic material available within landraces into breeding programs is not a sporadic event, but a permanent component of the breeding process.

Mixtures: the long-term approach

Pure-line selection within landraces is potentially dangerous because it tends to replace genetically heterogeneous populations such as landraces with genetically pure lines. The adoption by Syrian farmers of three different pure lines almost at the same time and in a relatively small geographical area — some farmers even adopted two different lines at the same time — suggests that the danger may be less dramatic than the spreading of single genotypes over very large areas, as in the case of HYVs. There is also evidence that in marginal environments, replacement of landraces is often only partial (Brush 1995). In principle, however, genetic uniformity contrasts with the genetic diversity characteristic of the agricultural systems of poor farmers in marginal areas. In these systems, diversity is preserved at one or more levels by using different crops on the same farm, different cultivars of the same crop, and heterogeneous cultivars. Diversity reduces the risk of crop failures due to abiotic and biotic stresses, while monoculture of a single genotype maximizes such risk.

One wonders why millennia of natural selection operating in harsh environments on a crop such as barley in the Fertile Crescent have left us with heterogeneous populations rather than with a single or few genotypes with superior adaptation. Perhaps yet another lesson that landraces are teaching is that it is the *structure* of the population, in addition to the genetic constitution of the individual components, that harbors the secret of adaptation to difficult and unpredictable environments (see next section). Constructing mixtures with a number of superior, yet genetically different, pure lines selected from landraces is the long-term objective of using landraces in the barley breeding program at ICARDA. This would provide the added benefit of a population buffering mechanism to the adaptation of the individual components (Grando and McGee 1990; Lenné and Smithson 1994). Though perhaps more time-consuming and experimentally more complex than the first two strategies, developing lines with the view of constructing mixtures is an additional way of responding to the need of poor farmers for stable yields.

Therefore, over the last 10 years we have conducted trials with mixtures of variable numbers of superior, yet genetically different, pure lines selected from landraces to compare yield and stability of pure lines and landraces. The results have shown not only the superiority of some specific mixtures, but also that some pure lines have yield and stability levels similar to those of mixtures. In the most recent of these trials we compared mixtures and pure lines within the two barley landraces, Arabi Abiad and Arabi Aswad, in a range of environmental conditions including the typical low-input stress-ful environments of farmers' fields in dry areas. The mixtures were made with either black-seeded or white-seeded lines. The black-seeded group had mixtures of 72, 34, 17, and 5 lines, the white-seeded group had mixtures of 75, 34, 15, and 5 lines. The constituent lines were either unselected (the more complex mixture) or derived from one (mixtures with 34 lines), two (mixtures with 17 and 15 lines), or three (mixtures with 5 lines) cycles of selection. The material was evaluated from 1990–1991 to 1994–1995 in 22 environments with mean yields ranging from 614 to 4385 kg/ha.

Linear regression analysis showed that black-seeded material tends to have lower average grain yield, lower response to higher yielding conditions, and higher frequency of positive intercepts than white-seeded material. In both groups the mixtures with five selected lines had an advantage over the more complex mixtures with unselected lines. In the black-seeded group (Table 3.12), the mixture of five lines had both average grain yield and regression coefficient significantly higher than the landrace Arabi Aswad with a slightly larger intercept, but did not have a clear advantage over the individual lines. In particular the line SLB 5-96 had a high average yield (2164 kg/ha), combined with a relatively good response (b = 0.97) and a positive intercept (a = 99.9). In the white-seeded group (Table 3.13), the mixture of five components had an advantage over the landrace Arabi Abiad with a higher intercept, and had an advantage over the single lines, com-bining a high average grain yield (2263 kg/ha) with a good response (b = 1.05) and positive intercept (a = 32.9). The only other line with a positive intercept (SLB 9-98) had a very low average grain yield (1833 kg/ha) and low response (b = 0.79).

The results suggest that the two Syrian barley landraces possess different buffering mechanisms. In the white-seeded group, which is less adapted to stress conditions, the advantage of the mixtures was more evident than in the more stress-adapted black-seeded group. The advantage of both five-component mixtures over the more heterogenous mixtures would indicate that yield stability may be achieved with a modest degree of heterogeneity,

Table 3.12 Average Grain Yield (kg/ha),
Regression Coefficient (b), and Intercept (a)
of Four Black-Seeded Mixtures, Five Lines,
and Three Checks

Material	Grain Yield	b	a
Mixtures			
MIXB 72	2017	0.88	147.4
MIXB 34	2060	0.96	10.2
MIXB 17	2076	0.97	11.8
MIXB 5	2131	0.93	150.1
Pure lines			
SLB 5-96	2164	0.97	99.9
SLB 5-07	2179	0.98	97.6
SLB 5-86	1950	0.84	167.0
SLB 5-31	2266	1.05	35.7
SLB 5-30	1982	0.86	144.2
Checks			
Arabi Aswad	1896	0.83	116.7
Tadmor	1971	0.86	140.0
Zanbaka	1946	0.84	154.9
$LSD_{0.05}$	164		

Table 3.13 Average Grain Yield (kg/ha),
Regression Coefficient (b), and Intercept (a) of
Four White-Seeded Mixtures, Five Lines, and
Three Checks

Material	Grain Yield	b	a
Mixtures			
MIXW 75	2237	1.15	−226.7
MIXW 34	2209	1.17	−277.1
MIXW 15	2174	1.08	−139.1
MIXW 5	2263	1.05	32.9
Pure lines			
SLB 9-63	2288	1.14	−144.4
SLB 9-71	2302	1.15	−152.3
SLB 9-76	2388	1.24	−248.5
SLB 9-09	2328	1.13	−86.8
SLB 9-98	1833	0.79	146.1
Checks			
Arabi Abiad	2202	1.14	−222.9
Arta	2414	1.19	−117.9
Harmal	2204	1.15	−248.8
$LSD_{0.05}$	164		

combined with the selection of superior lines. Like most studies on mixtures, these conclusions are strictly valid for the period under study. It may well be that the heterogeneity of the barley landraces from Syria has an advantage over longer periods of time than those usually covered by an experimental work. This is associated with the possibility, suggested by circumstantial evidence, of cross-pollination associated with an advantage of heterozygosity under drought (Einfeldt et al. 1996); this will determine continuous small adaptive changes in the genotypic composition of the landraces whose benefits can only be measured over longer periods of time than the 4 to 5 years of most experimental studies.

Understanding adaptation to stress

In addition to the contribution given to the breeding program, the landraces proved to be extremely useful experimental material for understanding adaptation to stress conditions in general, and the adaptive role of individual traits in particular. The genetic structure of landraces may be considered as an evolutionary approach to survival and performance under arid and semi-arid conditions (Schulze 1988). As indicated earlier, after millennia of cultivation under adverse conditions, natural and artificial selection have not been able to identify either an individual genotype possessing a key trait associated with superior performance or an individual genotype with a specific architecture of different traits. On the contrary, the combined effects of natural and artificial selection have led to an architecture of genotypes representing different combinations of traits. These populations can be extremely useful for understanding mechanisms that enhance stability in stress environments, not only from the population genetic point of view, but also for understanding the adaptive role of individual traits. In fact, although variable, landraces grown in environments characterized by a high frequency of stress conditions tend to present a high frequency of specific expressions of traits such as growth habit, cold tolerance, early growth vigor, and time to heading and maturity.

For example, barley lines extracted from landraces collected in five sites in the Syrian steppe (Table 3.14), compared with barley lines extracted from landraces collected in Jordan and with a wide range of modern barley genotypes, show a higher frequency of genotypes with prostrate or semi-prostrate growth habit, cold tolerance and short grain filling period, and a lower frequency of genotypes with good growth vigor and early heading. Their average grain yield in unfavorable conditions (Bouider 1989) was 984 kg/ha (ranging from 581 to 1394 kg/ha), more than twice the average grain yield of modern genotypes (483 kg/ha, ranging from crop failure to 1193 kg/ha). The average yield of the Syrian landraces in favorable conditions (3293 kg/ha) was 75% of the average yield of the modern germplasm in favorable conditions(4398 kg/ha). Although this particular set of data is based on one environment only, it confirms the existence of a trade-off between yield in

Table 3.14 Mean of Morphological and Developmental Traits[a] in 1041 Modern (Unrelated to Syrian or Jordanian Landraces) Barley Genotypes Compared with 322 Pure Lines Extracted from Syrian Landraces and 232 Pure Lines from Jordanian Landraces[b]

		Landraces	
Traits	Modern (n = 1041)	Syria (n = 322)	Jordan (n = 232)
1. Early growth vigor	2.5 b	3.2 a	2.4 b
2. Growth habit	2.8 c	4.0 a	3.1 b
3. Cold tolerance	3.0 a	1.3 c	2.3 b
4. Days to heading	117.9 b	121.2 a	116.9 c
5. Grain filling	39.3 a	35.5 c	37.4 b
6. YP	4398.0 a	3293.0 c	3947.0 b
7. YD	483.1 c	984.0 a	834.7 b

[a] Traits 1, 2, and 4–6 were scored or measured at Tel Hadya in 1987–1988 (504.2 mm rainfall), trait 3 was scored at Bouider in 1987–1988 (385.7 mm rainfall), and trait 7 was measured at Bouider in 1988–1989 (198 mm rainfall) on 521 modern lines, 92 Syrian landraces, and 86 Jordanian landraces. Means followed by the same letter are not significantly ($P < 0.05$) different based on t-test for samples of unequal size.

unfavorable conditions and yield in favorable conditions found in other sets of data based on a broader range of environments (Ceccarelli 1989).

Landraces collected in Jordan, from sites with milder winters than the Syrian steppe, have a higher frequency of genotypes with better early growth vigor, more erect habit, less cold tolerance, slightly longer grain filling period, and earlier heading than Syrian landraces. Their average grain yield in unfavorable conditions was only slightly lower (835 kg/ha) than Syrian landraces, while their average yield in favorable conditions (3947 kg/ha) was between that of the Syrian landraces and the modern germplasm. The highest yield of Syrian landraces under stress is not due to an escape mechanism, as they are the latest group in heading, and therefore could be a combination of resistance (or tolerance) and avoidance (prostrate habit and cold tolerance result in good ground cover) mechanisms.

Landraces are variable not only for above ground characteristics. A recent study (Table 3.15) shows that considerable variation exists for both the number and the length of seminal roots (Grando and Ceccarelli 1995) between different germplasm types. As mentioned earlier, seminal roots are important because in dry years they represent the only roots the plant produces. It appears that during the domestication of barley, the number of seminal roots has evolved from about three in *H. spontaneum* to five to seven in cultivated forms, while there has been a reduction in early root growth (root length) in modern varieties. In addition, the data show that for below ground characteristics — which are most likely important in relation to the use of water, one of the most limiting resources — there is

Table 3.15   Mean and Range of Variation for Number of
Seminal Roots and Their Maximum Length at Zadoks Stage
10 in Three Groups of Barley Germplasm

Germplasm Group	Number		Length	
	Mean	Range	Mean	Range
Modern	5.5	4.6–6.1	96.5	70.8–115.3
Landraces	5.1	4.4–5.9	118.8	107.4–131.6
H. spontaneum	3.3	3.0–3.8	107.3	97.2–118.1
LSD	0.7[a]	0.4[b]	14.9	11.6

[a] $LSD_{0.05}$ for group means comparison.

[b] $LSD_{0.05}$ for entry means comparison.

considerable variability within landraces. Therefore, the advantages of heterogeneity discussed in the previous section may apply underground as well as above ground.

The comparison between breeding lines with the highest yield under stress and those with the lowest yield under stress (Ceccarelli et al. 1991) indicates that the former were significantly earlier, more cold tolerant, had better ground cover and larger kernels, were taller under drought, and yielded less in favorable conditions. However, the range of variation for each of these traits in the genotypes with the highest yield under stress always overlaps with the range of variation of the same trait in the genotypes with the lowest yield under stress. This shows that the final performance (grain yield in unfavorable conditions) can be achieved by several combinations of a number of traits, and the role of each individual trait depends on the frequency, timing, duration, and severity of stresses, and on the type of stress. Therefore, it is probably the interaction among traits which plays a key role in determining the differences in overall performance rather than the expression of any single trait in isolation. Therefore, efforts to associate the superiority of landraces under stress conditions with specific traits and transfer them into modern varieties is unlikely to be successful. Long-term and sustainable improvements of yield stability should be based on population buffering, using mixtures of genotypes representing different, but equally successful, combinations of traits, as occurs in landraces.

Conclusions

This chapter has demonstrated that the utilization of the genetic variability within a collection of landraces from Syria and Jordan in a barley breeding program for the dry areas of the Fertile Crescent has been a success. This success is associated with the variability within landrace populations sampled at a given moment in the evolutionary process — a variability which could not be captured in a gene bank. To successfully use landraces in crop breeding for difficult environments, it is important to understand the value of landrace breeding programs, as well as the areas of research which need further exploration.

The value of landraces in plant breeding programs

The importance of landraces (and of wild relatives) for present and future breeding programs can be appreciated only if some conventional concepts in plant breeding, such as the need for widely adapted cultivars and the need to select under optimum conditions, are challenged. Because of their evolutionary history, landraces are useful as breeding material in stress environments and for poor farmers, in areas where years of conventional plant breeding have had virtually no impact. Landraces have a long history of specific adaptation to low-input agriculture. Under low-input conditions, landraces have maintained a considerable amount of genetic variability. The adaptation of landraces to specific soil and climatic conditions therefore results in the development of a diversity of improved varieties. Therefore, the conservation and use of landraces can contribute to increasing agricultural production without requiring additional inputs, as well as the conservation of biodiversity within crops.

In breeding a crop for difficult environments and poor farmers, selection (not only testing) must be conducted within the target environment and under the agronomic conditions of the local farmers. Research stations can be utilized for seed multiplication. If most breeding continues to be conducted under the high-input conditions of the research stations, landraces will have a limited value and *ex situ* collections will continue to be poorly utilized.

There is, however, an implicit danger that a breeding approach based on the use of landraces may eventually accelerate the rate of genetic erosion. As indicated earlier, the success that is likely to occur by exploiting the variability within landraces through pure-line selection may lead to the widespread adoption of the pure lines and the disappearance of the landraces. One approach to prevent the replacement of landraces is that of participatory plant breeding. In implementing a participatory plant breeding program where farmers select from a wide range of germplasm present in their own fields, we have found that farmers select material derived from landraces more frequently than other material. Farmers also want to know the nature and origin of the material they select, particularly that which is performing well. Their understanding that the landraces they have grown over long periods of time are capable of continuously generating new types which can improve the living standards of present and future generations could become a key factor in promoting their interest in conserving the original landrace, while adopting new lines and mixtures. Therefore, participatory plant breeding could generate considerable farmer interest in *in situ* conservation.

Research areas for further exploration

Although successful, the work on landraces described in this chapter leaves a number of questions unanswered. For example, we still ignore the complexity of the population structure of landraces both in terms of the number of different homozygotes (since barley is a self-pollinated crop) and the occurrence and frequency of heterozygotes due to natural cross-pollination. Similarly, we do not know if there is any geneflow between *H. spontaneum* and cultivated barley. If geneflow does occur, how frequent is it, and what role does this introgression play in the performance and stability of the landraces?

We have no systematic description of the nature and amount of genetic variability available in landraces in different geographical areas, of the spectrum of adaptation of populations collected in different areas, and of the frequency of useful traits. Yet, such information is essential to define the areas of adaptation of different landraces (which would be useful where germplasm is lost in specific areas due to natural or political calamities), and to identify priority areas for *in situ* conservation.

One research area that requires major emphasis is that of mixtures. In some crops (including barley), the release of mixtures as cultivars can be done even in the presence of the restrictive regulations on variety release and seed certification, because virtually all seed comes from the informal seed system. By evaluating bulk samples with farmers' participation, successful bulks can find their way directly into the informal seed system. For other crops, years of breeding for uniformity have generated the widespread aversion of breeders to heterogeneity, which contrasts dramatically with the heterogeneity of the material best adapted to difficult environments. The type of mixtures that should be investigated depends on the context in which the crop is grown. In the case of barley in the Fertile Crescent, for example, it might be necessary to consider the presence of *H. spontaneum* if it can be shown that there is indeed geneflow between wild and cultivated barley.

Landraces are adapted to their environment, and they fit into the farming systems of their area of adaptation. They are often essential components in the diet, and in many cases they are the only food or feed available. The welfare of people depending on landraces should and can be improved not by replacing landraces but by improving them. Maintaining the genes of landraces in breeding programs and through *in situ* conservation programs is a moral obligation toward those many farmers who have maintained landraces over millennia.

Acknowledgment

The work on barley landraces conducted in Syria after 1987 has been supported by the government of Italy.

References

Asfaw, Z. 1989. Variation in hordein polypeptide pattern within Ethiopian barley *Hordeum vulgare* L. (*Poaceae*), *Hereditas* 110:185–191.

Blum, A., G. Golan, and J. Mayer. 1991. Progress achieved by breeding open-pollinated cultivars as compared with landraces of sorghum, *Journal of Agricultural Science* 117:307–312.

Blum, A., G. Golan, J. Mayer, B. Sinmena, L. Shpiler, and J. Burra. 1989. The drought response of landraces of wheat from the northern Negev Desert in Israel, *Euphytica* 43:87–96.

Brown, A.H.D. 1978. Isozymes, plant population genetic structure, and genetic conservation, *Theoretical and Applied Genetics* 52:145–157.

Brown, A.H.D. 1979. Enzyme polymorphism in plant populations, *Theoretical Population Biology* 15:1–42.

Brush, S.B. 1995. *In situ* conservation of landraces in centers of crop diversity, *Crop Science* 35:346–354.

Ceccarelli, S. 1984. Utilization of landraces and *H. spontaneum* in barley breeding for dry areas, *Rachis* 3(2):8–11.

Ceccarelli, S. 1989. Wide adaptation: how wide? *Euphytica* 40:197–205.

Ceccarelli, S. 1996. Adaptation to low/high input cultivation, *Euphytica* 92:203–214.

Ceccarelli, S., E. Acevedo, and S. Grando. 1991. Breeding for yield stability in unpredictable environments: single traits, interaction between traits, and architecture of genotypes, *Euphytica* 56:169–185.

Ceccarelli, S. and S. Grando. 1996. Drought as a challenge for the plant breeder, *Plant Growth Regulation* 20:149–155.

Ceccarelli, S., S. Grando, and J.A.G. van Leur. 1987. Genetic diversity in barley landraces from Syria and Jordan, *Euphytica* 36:389–405.

Damania, A.B., M.T. Jackson, and E. Porceddu. 1985. Variation in wheat and barley landraces from Nepal and the Yemen Arab Republic, *Z. Pflanzenzuchtung* 94:13–24.

Damania, A. and E. Porceddu. 1983. Variation in landraces of Turgidum and Bread Wheats and sampling strategies for collecting wheat genetic resources. *Proceedings of the 6th International Wheat Genetic Symposium*. Kyoto, Japan (28 Nov.– 3 Dec., 1983), Plant Germplasm Institute.

Einfeldt, C.H.P., S. Ceccarelli, S. Gland-Zwerger, and H.H. Geiger. 1996. Influence of heterogeneity and heterozygosity on yield under drought stress conditions. *Proceedings of the V International Oat Conference and of the VII International Barley Genetics Symposium*. A. Slinkard, G. Scoles, and B. Rossnagel (eds.). Saskatoon, Canada (July 30–Aug. 6, 1996), University Extension Press, University of Saskatchewan.

Elings, A. and M.M. Nachit. 1991. Durum wheat landraces from Syria. I. Agroecological and morphological characterization, *Euphytica* 53:211–224.

Erskine, W. and N.A. Choudhary. 1986. Variation between and within lentil landraces from Yemen Arab Republic, *Euphytica* 35:695–700.

Grando, S. and S. Ceccarelli. 1995. Seminal root morphology and coleoptile length in wild (*Hordeum vulgare* ssp. *spontaneum*) and cultivated (*Hordeum vulgare* ssp. *vulgare*) barley, *Euphytica* 86:73–80.

Grando, S. and R.J. McGee. 1990. Utilization of barley landraces in a breeding program. In *Biotic Stresses of Barley in Arid and Semi-Arid Environments*. Bozeman: Montana State University.

Harlan, J.R. and D. Zohary. 1966. Distribution of wild wheats and barley, *Science* 153:1074–1080.

Lagudah, E.S., R.G. Flood, and G.M. Halloran. 1987. Variation in high molecular weight glutenin subunits in landraces of hexaploid wheat from Afghanistan, *Euphytica* 36:3–9.

Lakew B., Y. Semeane, F. Alemayehu, H. Gebre, S. Grando, J. van Leur, and S. Ceccarelli. 1997. Exploiting the diversity of barley landraces in Ethiopia, *Genetic Resources and Crop Evolution* 44:109–116.

Lenné, J.M. and J.B. Smithson. 1994. Varietal mixtures: a viable strategy for sustainable productivity in subsistence agriculture? *Aspects of Applied Biology* 39:163–172.

Martin, G.B. and W.M. Adams. 1987a. Landraces of *Phaseolus vulgaris* (Fabaceae) in northern Malawi. I. regional variation, *Economic Botany* 41:190–203.

Martin, G.B. and W.M. Adams. 1987b. Landraces of *Phaseolus vulgaris* (Fabaceae) in northern Malawi. II. generation and maintenance of variability, *Economic Botany* 41:204–215.

Porceddu, E. and G.T. Scarascia Mugnozza. 1983. Genetic variation in durum wheat, *Proceedings of the 6th International Wheat Genetic Symposium*. Kyoto, Japan (28 Nov.–3 Dec., 1983), Plant Germplasm Institute.

Schulze, E.D. 1988. Adaptation mechanisms of non cultivated arid-zone plants: useful lesson for agriculture? In *Drought Research Priorities for the Dryland Tropics*, F.R. Bidinger and C. Johansen (eds.). Patancheru, India: ICRISAT.

Spagnoletti-Zeuli, P.L., C. De Pace, and E. Porceddu. 1984. Variation in durum wheat populations from three geographical origins. I. material and spike characteristics, *Euphytica* 33:563–575.

van Leur, J.A.G., S. Ceccarelli, and S. Grando. 1989. Diversity for disease resistance in barley landraces from Syria and Jordan, *Plant Breeding* 103(4):324–335.

Weltzien, E. 1982. Observation on the growth habit of Syrian and Jordanian landraces of barley, *Rachis* 1:6–7.

Weltzien, E. 1988. Evaluation of barley (*Hordeum vulgare* L.) landraces populations originating from different growing regions in the Near East, *Plant Breeding* 101:95–106.

Weltzien, E. 1989. Differentiation among barley landrace populations from the Near East, *Euphytica* 43:29–39.

Weltzien, E. and G. Fischbeck. 1990. Performance and variability of local barley landraces in Near-Eastern environments, *Plant Breeding* 104:58–67.

chapter four

The barleys of Ethiopia

Zemede Asfaw

Introduction

Recognized as one of the world's most ancient food crops, barley has been an important cereal crop since the early stages of agricultural innovations 8,000 to 10,000 years ago. Throughout history, barley has undergone continuous manipulation in an effort to optimize its use for human consumption and as animal feed. Barley has been used as a model organism in experimental botany, the plant of choice because of its short life cycle and morphological, physiological, and genetic characteristics. Globally, barley ranks fourth among cereal crops in both yield and acreage, after wheat, rice, and maize (Munck 1992b). With advances in food production and agriculture, major dietary shifts from barley to rice and/or wheat have resulted in the decline in barley consumption, with the exception of societies — particularly those relying on traditional, small-scale agricultural systems — in which its use as human food has continued to the present.

The world has now "re-discovered" barley as a food grain with desirable nutritional composition including some medicinal properties. Barley breakfast foods and snacks are increasingly available, driven by recent research findings, which show that barley fiber contains beta-glucans and tocotrinols, chemical agents known to lower serum cholesterol levels (Burger et al. 1981; Anderson et al. 1991). In Ethiopia, barley is the third most important cereal crop next to teff and maize. It is the staple food grain for Ethiopian highlanders, who manage the crop with indigenous technologies and utilize different parts of the plant for different purposes.

Efforts to improve barley have demonstrated a preference for a limited number of modern, genetically uniform cultivars suited for high input agriculture, to the neglect of the various farmers' varieties, or landraces, on which a large sector of the human population has subsisted for millennia. The trend has narrowed the genetic base of the local material, leading to the gradual replacement of landraces with modern barley cultivars or of other crops such as wheat and oats. One consequence of this replacement is the loss of indigenous knowledge associated with replaced landraces. It is noted that some earlier morphotypes of Ethiopian barley (e.g., hooded barley; Bell 1965) are no longer found in cultivation. Some Ethiopian barley types (e.g., smooth awned types, hull-less types) kept at the Gatersleben gene bank in Germany (Index Seminum 1983) are not found in the country at present. Some varieties reported as abundant during the Vavilovian expedition (many naked and some rare covered forms) (Orlov 1929) could not be found in those areas (Asfaw 1988). The global trend has been to select for a few high yielding types, thus narrowing the genetic base of a crop. This trend has influenced the direction of Ethiopia's limited barley research over the past four decades. In crop genetic resources conservation efforts, Ethiopian barley has been identified as a priority crop since the 1920s, and extensive germplasm collections have been deposited in gene banks all over the world, especially in Russia and the U.S. (Orlov 1929; Ciferri 1940, 1944; Negassa 1985). Both the usefulness of barley and its high genetic and morphological diversity have rendered barley conservation a matter of top priority. This is evidenced by a long history of conservation in gene banks around the world since the 1920s, beginning formally in Ethiopia in 1976. *Ex situ* germplasm conservation has facilitated the preservation of the diversity present at a given point in time, but does not preserve the dynamic co-evolutionary processes that take place when landraces are continuously cultivated in their natural agroecological settings. To remedy this shortcoming, the need for complementary *in situ* conservation has been recommended and is under serious consideration (Feyissa 1995; Soleri and Smith 1995; Altieri and Montecinos 1993).

Scientists are currently working to improve barley using genetic engineering and other modern techniques; they are looking forward to the formulation of barley ice cream and many other fabulous products for future markets. Another area of research concentrates on alternative approaches for sustainable use and conservation of the diversity in the barley gene pool. This approach focuses on *in situ* conservation of barley landraces — a new line of thought rooted in the traditional practices that have preserved the indigenous farmers' varieties. Traditional farming systems have the dual functions of production and conservation since the entire agroecosystems are crop germplasm repositories (Altieri and Montecinos 1993). This chapter highlights the case of barley in Ethiopia, focusing on the importance of traditional management and cultural practices associated with the landraces. Traditional farmer practices are viewed in the light of on-farm conservation activities being implemented under a new landrace on-farm conservation project, *A Dynamic Farmer-Based Approach to the Conservation of Ethiopia's Plant Genetic Resources,* supported by the Global Environment Facility and implemented by the Biodiversity Institute of Ethiopia in collaboration with other institutions.

The barley crop

General botany, phylogenetic relations, and classification

Barley belongs to the genus *Hordeum* L. in the tribe Triticeae of the family Poaceae. The genus *Hordeum* is a distinct genus in the tribe, well distinguished by three one-flowered spikelets at each rachis node. Its taxonomy and phylogeny have been studied by many scholars including Orlov and Åberg (1941), and von Bothmer et al. (1981). Believed to have differentiated from *Agropyron-Elymuss*-like ancestors (von Bothmer et al. 1981), the barley genus, *Hordeum*, is a relatively small genus with about 28 species distributed over wide geographical areas and diverse ecological habitats. Its three main centers of distribution are southern South America, western North America, and southwestern to central Asia. Species occurring in the Americas, Eurasia, the Mediterranean–Middle East, and Africa number 19, 5, 3, and 1, respectively. The greatest diversity of the genus is found in southern South America, which together with southwestern Asia constitutes the primary centers of diversity (von Bothmer et al. 1981). The two areas of its primary center are connected by a single endemic species *(Hordeum capense)* found in South Africa.

Two parallel hypotheses have been posited to explain an ancient differentiation of the genus: one proposes that ancient forms of *Hordeum* were distributed in a larger area including South America and southern and eastern Africa up to central Asia; a second hypothesis asserts that early migrations of the genus took place in one primary center, most likely South America, migrating to Asia via South Africa. The former view, which advocates a wider initial distribution of the barley genus, is considered more plausible (von Bothmer et al. 1981), based on the fact that there is at least one primitive group in each major area. According to modern treatment, *Hordeum vulgare* L. is differentiated into two subspecies: *spontaneum* and *vulgare*. The former subspecies contains all the *spontaneum* group and is the immediate ancestor of all cultivated types. All of the cultivated types are lumped into subspecies *vulgare*. The main feature distinguishing between the two subspecies is that the *spontaneum* types have brittle rachis while the *vulgare* types have tough rachis. The *spontaneum* group is believed to have been derived from the wild *Hordeum* species, *H. murinum* and *H. bulbosum*, characterized by well-developed lateral florets (von Bothmer and Jacobsen 1985). A scheme for the taxonomy and classification of the cultivated and *spontaneum* groups has been developed by Orlov and Åberg (1941). The cultivated group is frequently treated in a taxonomic scheme consisting of convarieties — multiple varieties having the same or equal taxonomic status and displaying discernible morphologies and generally recognized as distinct cultivated varieties/forms (Table 4.1). The convarity category is the botanical equivalent of cultivar groups.

Table 4.1   Infraspecific Taxonomic Groups of Barley
(Grillot 1959; von Bothmer et al. 1981)

1.	Convar. **Vulgare**	all rachis spikelets fertile, awns long
	Var. *Vulgare*	caryopsis not naked, spike lax
	f. *Vulgare*	rachis tough
	f. *Agriochathon*	rachis brittle (a wild 6-row form)
	Var. *Coeleste*	caryopsis naked, spike lax
	f. *Coeleste*	awned
	f. *Trifurcatum*	awns bifurcate
	Var. *Hexastichon*	caryopsis not naked, spike compact
	Var. *Revelatum*	caryopsis naked, spike compact
2.	Convar. **Distichon**	fertile central spikelets, sterile or male fertile laterals
	Var. *Distichon*	caryopsis not naked (covered)
	Var. *Nudum*	caryopsis naked
	Var. *Zoecnthon*	caryopsis not naked, spikes short and broad, awns divergent
	Var. *Deficiens*	caryopsis not naked, laterals glume-like appendages
3.	Convar. **Intermedium**	lax type of six row
4.	Convar. **Labile**	irregular spike row number

History of cultivation and use

The earliest cultivation of barley is believed to have begun some 8,000 to 10,000 years ago in the area of the Middle East known as the Fertile Crescent (Giles and von Bothmer 1985; von Bothmer and Jacobsen 1985). This conclusion, still debated by many (e.g., Bekele 1983; Negassa 1985), is based on archaeological findings and the presence of *spontaneum* types both in the absolute wild state and as weeds in crop fields in southeast Asia. The spontaneous form also occurs as weed in North Africa, probably harvested in prehistoric times from wild stands as far south as the Nile Valley of Egypt (Wendorf et al. 1979). The crop is now grown worldwide with greater concentration in temperate areas and high altitudes of the tropics and subtropics. The greatest diversity of barley in terms of morphological types, genetic races, disease-resistant lines, and endemic morphotypes exists in Ethiopia (Orlov 1929; Huffnagel 1961).

Initially one of the dominant food grains, barley has been surpassed by rice and wheat in many countries. In traditional societies barley continues to be a very important food grain. Internationally, its importance as a feed and brewing grain has increased through the years. Recent findings on the nutritional qualities of barley have begun to make it a desirable food item even in those countries where its consumption had declined for many years (Anderson et al. 1991). It is likely that traditional barley landraces will attract the consumer society as they the tend to be more nutritionally balanced than modern varieties. With increasing consumer awareness of nutritional composition of diets, landraces are anticipated to fetch higher market prices. It may not be too long before the genotypic attributes of a crop begin to positively influence its market price. The potential for some barley landraces in this regard appears to be high (e.g., some naked and partially naked types).

Barley in Ethiopia

Antiquity and botanical affinities

The persistence of two parallel hypotheses regarding the possible origin of barley in Ethiopia has resulted in a lively debate among crop scientists. A number of studies have dealt with various aspects of the debate, including crop domestication patterns in Africa (Porteres 1976; Purseglove 1976), interpretations of linguistic evidence (Ehret 1979), and archaeological and historical documentation and analyses (Brandt 1984). Sources agree to the extent that barley has been in cultivation in Ethiopia for at least the past 5,000 years, based on evidence that it was cultivated about 3000 B.C. by the Agew people of northwest Ethiopia (Gamst 1969). In parts of southern and central Ethiopia, the history of barley cultivation is reported to have coincided with the history of the plow culture. It is said that barley was considered a sacred crop by the Oromo people of southern Ethiopia (Haberland 1963). Bekele (1983) challenged the single origin hypothesis, arguing that based on barley flavonoid data, it is highly plausible that *spontaneum* gene was introgressed and gradually swamped up into the *vulgare* gene pool in Ethiopia.

Barley researchers have long considered the Ethiopian barley stock as an isolated line that evolved independently from the mainstream of world barley evolution, posited to be around southwest Asia (Harlan 1968). Such claims were based on a limited number of experimental results that gave clues of partial sterility and reduced seed set ratios of crosses between Ethiopian barley and those from Europe and Asia (Smith 1951; Jonassen and Munck 1981). These comments prompted a major question as to how far the Ethiopian barley gene pool has differentiated from that of its wild ancestor. A reciprocate crossing experiment undertaken between *spontaneum* lines and a selection from the Ethiopian *vulgare* showed high levels of hybrid viability and fertility (Figure 4.1), assessed on the basies of pollen fertility, seed set, hybrid viability, and vigor (Asfaw 1991; Asfaw and von Bothmer 1990). Mechanisms of character inheritance were easily followed as they conformed to the known ratios, demonstrating the ease with which genes can be transferred between *spontaneum* and the Ethiopian stock. The free intercrossing of the two subspecies has been reaffirmed (von Bothmer and Seberg 1995).

History of exploration and studies

Foreign crop exploration missions began in Ethiopia 400 years ago, at which time barley was a primary crop under investigation. Early travelers, including the Portuguese Francisco Alvares, who explored Ethiopia in the 1520s (Alvares 1961), have recorded the wide occurrence of barley. The presence of domestic varieties of barley in Ethiopia was registered by many 19th

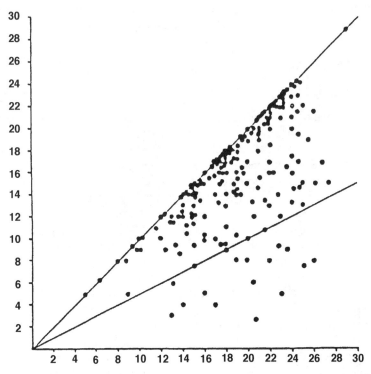

Figure 4.1   Fertility between *spontaneum* and Ethopian barley types [Florets (*vertical*), Grains (*horizontal*)]. Points on the diagonal line show F2 generation plants in which all florets set seeds and the second line marks the level where half of the florets set seeds and the other half are aborted (Asfaw and von Bothmer 1990).

century crop taxonomists, including Kornicke and Atterberg (Orlov 1929), and the first scientific botanical account was given by Chiovenda (1912). Later studies noted the unique features of the barley cultigen grown in Ethiopia (Orlov 1929; Ciferri 1940, 1944; Vavilov 1951). Barley was also targeted in the germplasm exploration studies of American and British missions (U.S. Operation Mission to Ethiopia 1954; Huffnagel 1961). Judging from the content and emphasis of the descriptions produced, the early explorers appeared to have been most attracted by the morphological variation and the endemic types as reported subsequently (Orlov 1929). The early studies covered aspects of the morphology, agronomy, ecology, diversity, evolution, genetics, and taxonomy of the barley grown in Ethiopia. The more comprehensive studies were those of Russian and Italian investigators (Orlov 1929; Ciferri 1940, 1944), who made field explorations and collections in Ethiopia, as well as observations through cultivation experiments and laboratory analyses in their respective countries. The methods and results from these studies were not made available locally and not taken up by resident researchers.

As international researchers increasingly realized the potential of Ethiopia's diverse barley types, particularly for disease resistance, interest shifted toward the utilization of germplasm in the breeding and development of modern cultivars. Many reputed modern barley cultivars in Europe and the U.S. owe their resistant genes to material originally collected from Ethiopia (Hoyt 1988). Harlan (1968, 1969) publicized the view that Ethiopian barley types were not favored for improvement as modern cultivars, while emphasizing their immense value as gene donors for barley improvement. Although this claim was based on observations made when the material was grown far away from its natural habitat and geographical range, it seems to have influenced the direction of barley research in Ethiopia. Other researchers have fully acknowledged the attractive traits of Ethiopian landraces including large kernel size, high tillering, and large 1000-grain weight (Orlov 1929; Huffnagel 1961; Westphal 1975), and favorable nutritional qualities such as higher protein/lysine content (Munck 1992b; Jonassen and Munck 1981) and cholesterol-reducing chemical agents (Anderson et al. 1991; Heen et al. 1991). More recent studies have focused on the resistance of the Ethiopian types to known pathogens, germplasm conservation and utilization, assessment of diversity, and biological gene markers (Qualset 1975; Metcalfe et al. 1978; Bekele 1983; Negassa 1985; Engels 1986). Asfaw has shown the wide diversity in morphological characters (1988, 1990) and hordein polypeptide pattern (1989c), and the potential for wide hybridization (Asfaw and von Bothmer 1990). Demissie (1996) investigated morphological and molecular diversity markers and stressed the implications for *in situ* and *ex situ* conservation. Other studies have identified a wealth of ethnobotanical knowledge associated with barley landraces in Ethiopia (Asfaw 1990). Ethiopian barley types have contributed significantly to the understanding of barley, increasing its status as a soundly fathomed crop on a worldwide scale.

Distribution throughout Ethiopia

Barley is cultivated in every region of Ethiopia and demonstrates wide ecological plasticity and physiological amplitude throughout the country (Asfaw 1988, 1989; Lakew et al. 1996). The crop is cultivated from 1,400 to over 4,000 meters above sea level, with the greatest frequency and diversity occurring between 2,400 and 3,400 meters in the northern and central regions of Ethiopia (Figure 4.2). Diverse landraces and morphological classes of barley are adapted to specific sets of agroecological and microclimatic regimes throughout the country. The higher preponderance of some morphotypes (six-rows, naked caryopsis types, dense spikes, higher anthocyanic types) and some hordein polypeptide patterns at higher altitudes, other types (e.g., two-rows, lax types), and other hordein polypeptide patterns at lower elevations are documented (Asfaw 1988, 1989b). Differential distribution, including abundance of primitive flavonoid patterns (Bekele 1983), resistant genes (Negassa 1985), and phenotypes and diverse molecular

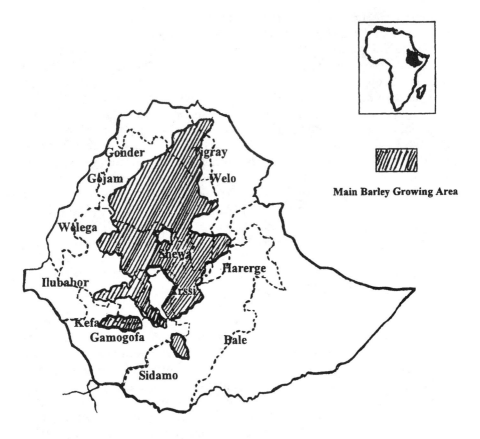

Figure 4.2   Distribution of Barley in Ethiopia.

markers have been reported (Asfaw 1988, 1989c; Demissie 1996; Demissie and Bjornstad 1996, 1997). Within the general barley growing areas and the optimal agroecologic range there are pockets in which are concentrated some morphological and chemical groups that can guide future conservation strategies. On the whole the southern and southeastern highlands harbor more morphotypes than the central and northern highlands. However, some individual localities within both zones (e.g., Kembata, Galessa-Tululencha, Chencha) are recognized as pockets of higher number of morphotypes per field, as illustrated by a study carried out in Jibat and Mecha (Asfaw 1990) revealing higher number of morphotypes per field and in the entire locality. While some barley morphotypes are widely distributed, others are restricted to narrow ranges and isolated pockets. Some types are still sheltered from the direct effects of invading modern agrotechnology such as the use of modern cultivars, inorganic fertilizers and pesticides as they are found in places not easily accessible except to the owners. Hence, Ethiopia is a promising site for both *ex situ* and *in situ* conservation of barley. Demissie and Bjornstad (1996, 1997) recommended that collection and conservation of barley germplasm in Ethiopia should take account of the differential distribution of polymorphism in phenotypes, isozymes, and hordein genotypes.

Significance and modes of consumption

As the third most important cereal crop cultivated in Ethiopia, barley is grown primarily for local food and beverage consumption. For small-scale highland farmers, barley is the predominant subsistence crop. It is typically produced two times per year, during the long and short rainy seasons that extend from June to September and from February to April, respectively. In some regions, barley is also produced three times per year, drawing on residual moisture supplemented with irrigation. While some landraces are cultivated during both primary growing seasons, others are adapted only to the long rainy season. In terms of consumption, Ethiopia ranks second only to Morocco with respect to the number of kilograms (68 and 19, respectively) of barley consumed per person per annum (FAO 1990, in Bhatty 1992). Whereas barley consumption declined in many countries, it continued at the same level in Ethiopia, where nearly 40% of the total grain produced is used as food (Gebre and Pinto 1977).

Within Ethiopia, the highest levels of barley consumption occur in highland areas where it is widely cultivated, accounting for the bulk of the total crop harvest. In these areas barley consumption begins at the milky stage of grain maturation when youngsters remove the awns from the green unripe spikes, crush them between the palms, blow away the fragments of the rachis and glumes, and consume the tasty raw green grains in the field in limited quantities. Such unripe spikes may also be green-roasted over fire. Similarly, a sheaf of ripe barley can be roasted in the fire, crushed between the palms and the grains eaten as a supplementary or "waiting" food. Different kinds of bread, dough balls, porridge, soup, and gruel are made in every household from any barley type, but there are preferred types for different methods of preparation (Asfaw 1990). Many alcoholic and nonalcoholic local beverages are brewed in the household from barley grains for daily consumption or for holidays and celebrations. The barley straw is used in the construction of traditional huts and grain stores either as thatching or as a mud plaster (Figure 4.3). The barley crop-residue is used as fodder mainly for bovine cattle and equine. The small grains that fail to fill up and those crushed in the process of threshing and consequently mix with the chaff are kept aside for chicken feed (and sometimes small ruminants and riding horses or mules) by some families. Some barley types are purposely cultivated for their special uses (e.g., partially naked types for roasted grains) while many others are more of multipurpose types.

Special features of barley in Ethiopia

The cultivated forms of *Hordeum* are a group of interfertile lines distinguished by differences in spike characters. More than 180 botanical forms of

Figure 4.3   Barley stalks constructed in the traditional way (Shewa).

barley, represented by a larger number of agricultural "varieties," occur throughout the world (Bell 1965). A large number of botanical varieties have been recognized (e.g., Orlov 1929; Orlov and Åberg 1941; Ciferri 1944). Giessen et al. (cited in Huffnagel 1961) are reported to have identified 170 types from Ethiopia, grouped into five convarieties (viz. Convar. *Deficiens, Distichon, Hexastichon, Intermedium,* and *Labile*) (see Table 4.1). The *Deficiens* and *Labile* forms are endemic to Ethiopia. Their occurrence only in Ethiopia supports the hypothesis that a unique evolutionary reduction in the morphological characters of barley occurred in Ethiopia; in fact, the *Labile* group represents an intermediate form. This observation, together with the view that the barley genus had enjoyed a wider distribution in the geologic past, including in eastern Africa, supports the argument that barley probably originated independently in Ethiopia as well (Bekele 1983; Negassa 1985). Recent studies also brought to light the presence of a large number of botanical forms and morphological types of barley (>60) and hordein groups (>40) in the Ethiopian barley material (Asfaw 1988, 1989c; Demissie 1996).

The main groups can be classified as hulled, hull-less, and partially hulled types with six-row, two-row, and irregular morphologies, and varied spike shape, density, and pigmentation. These distinct characteristics are further combined with glume and lemma characters that display a wide range of variation in size, shape, color, and texture. The wide diversity is further accentuated by the coexistence of features considered primitive in cultivated barley, such as covered caryopsis, bigger plants, pubescence, well-developed glumes and anthocyanic straws, with more advanced features, including short awns, large grains, deficient forms, straw yellow spikes and grains, and naked types. All the convarieties, varieties, and forms listed in Table 4.1, except the form *agriochrithon*, occur in Ethiopia.

The more common botanical forms of Ethiopian barley are *Deficiens, Pallidum, Nutans,* and *Nigrum* types. Some of the morphological types are reported to be endemic to Ethiopia (Orlov 1929; Harlan 1969). The pyramidal, parallel, and hull-less types are restricted in distribution and are less abundant, although mobility between regions is a possibility as evident from some vernacular names of barley landraces.

The frequency of six-row types, hull-less types and those with compact and colored spikes is highest at higher altitudes. The dominant barley types vary between fields, localities, and regions, with up to 12 distinct morphotypes present in a single barley field that averages 0.5 ha. Since morphological variations expressed under uniform ecological conditions are likely to be genetic, the different morphotypes seen in a field are best considered manifestations of gene differences (Figure 4.4). Farmers' knowledge of this diversity survives in the elaborate folk taxonomy and system of nomenclature as well as in the beliefs, value systems, cultural songs, and aphorisms (Asfaw 1990, 1996). The extent to which value systems and expressions of culture reflect upon farmers' knowledge of crop diversity has been documented for other important major crops of the world in their centers of domestication (Bellon 1996).

The main popular groups of barley in Ethiopia

Barley is usually grouped into morphological categories based on spike row number at the top level to form major categories. However, it is observed that farmers and communities more frequently use barley groups based on differences of caryopsis or kernel type. For routine application caryopsis type is easily understood as it is a utilitarian criterion. Whether spike row number or caryopsis type is used at the top level, other spike characters have to be used for complete identification of the barley. The top level gives only major classes of barley. Using caryopsis type, three main barley groups are easily distinguished: hulled, hull-less, and partially hulled. This system is very efficient to apply at the house level and frequently used by women in both rural and urban areas. While classifying Ethiopian barley using morphological characters, application of clustering technique (ordination) gave a distinct group of only naked types at initial classification also revealing that the character is also botanically distinct and more conclusive than row number (Asfaw 1988). Caryopsis type is one character that is used for barley classification both under the traditional and the modern systems that is easier for routine application. Formal taxonomy (see Orlov and Åberg 1941; Grillot 1959) begins with spike row number, but also uses caryopsis type as one of the essential criteria since it is a distinct character on the spike. If, however, caryopsis type is used at the top level, barley types can be easily categorized into three major groups as hulled (covered), hull-less (naked), and partially hulled types. Each of these can then be further classified using other characters given in Table 4.2. The three major groups of barley based on differences in caryopsis type are highlighted below.

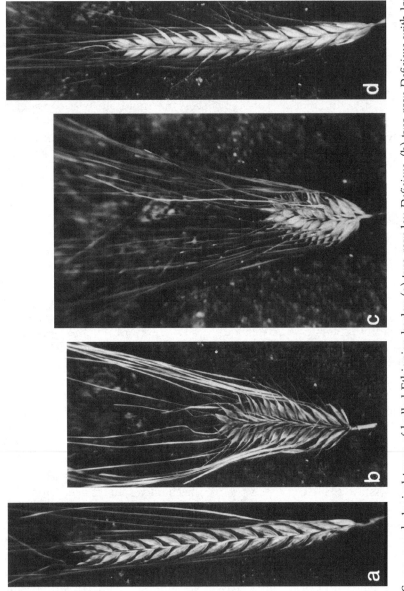

Figure 4.4   Some morphological types of hulled Ethiopian barley (a) two-row lax *Deficiens*, (b) two-row *Deficiens* with long and broad outer glumes and diverging awns; (c) six-row, dense spike; (d) two-row, lax spike, *Deficiens* type with broad outer glumes.

Table 4.2 Spike Characters Used in Folk and Modern Classification (combination not encountered) (Asfaw 1996)

Character	Six-Row	Two-Row *Deficiens*	Two-Row *Nutans*	Irregular
Spike	dense	dense	dense	—
	lax	lax	lax	lax
	long	long	long	long
	short	—	short	—
	stout	—	—	—
Kernel/caryopsis	hulled	hulled	Hulled	hulled
	hull-less	—	hull-less	—
	hull-partial	partial	partial	—
	white	white	white	white
	black	black	black	black
	purple	purple	purple	purple
Appendages	hood	—	—	—
	awn	awn	awn	awn
	awn long	long	long	long
	awn short	short	—	—
	awn rough	rough	rough	rough
	awn smooth	—	—	—
	awn diverging	diverging	diverging	—
	awn converging	converging	converging	—
	awn persistent	persistent	persistent	persistent
	awn brittle	—	—	—
Outer glumes	broad	broad	—	—
	narrow	narrow	narrow	narrow
	long-awn	—	—	—
	short-awn	short-awn	short-awn	short-awn

Hulled barley. This group is known as the farmer's "true" barley. The husk adheres to the grain, requiring an arduous dehulling process to make the grain suitable for consumption. It is the largest group in terms of cultivated area, the provenance, and the number of morphological types (see Figure 4.5). All hulled barley, including partially hulled types, accounts for about 70% of the morphologically distinct barley types in Ethiopia. Hulled barley is the most diverse major category including six-rows, two-rows, irregular forms, dense, lax, hooded, long and short awned, rough and smooth awned types (Table 4.1). Traditional farmers in Ethiopia consider this group less labor-intensive in the field and of a relatively higher grain yield than other barley types. In terms of food preparation, however, hulled barley is less desirable as it is extremely time and labor intensive as reported by women.

Hull-less barley. In the hull-less (naked) barley group, the husk falls free from the grain upon threshing. The hull-less type of Ethiopian barley constitutes the genetic pool from which the lysine-high protein, hiproly

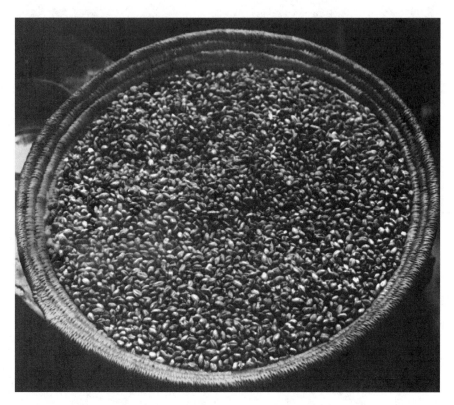

Figure 4.5 Roasted barley brought for selling at bus stop (Sheno town, Shewa).

barley was recovered by screening (Munck 1992b; Jonassen and Munck 1981). Included are two-row, six-row, lax, and dense forms (see Table 4.2). Absence of the hull is a recessive character encountered in six- and two-row *Nutans* types; the character has not been encountered in *Deficiens* and *Labile* forms.

Throughout Ethiopia, the frequency of hull-less barley is low and the distribution is restricted to the highland regions of Shewa, Gonder, and Tigray. Most of the morphotypes occur as rare mixtures among fields of hulled and partially hulled types; few pure stands have been documented. In one locality in Shewa (Jibat), where the highest concentration of hull-less types was found, a total of 31 distinct barley types were identified, 4 (12%) of which were of the hull-less type (Asfaw 1990). Early surveys found that hull-less types constituted a substantial amount of the barley grown in Ethiopia, and noted a great diversity within the hull-less types (Orlov 1929; Ciferri 1944). Ciferri (1944) found that, throughout Ethiopia, hull-less types accounted for 38% of cultivated barley; Orlov (1929) recorded hull-less types as 36% of the total in the Addis Ababa region.

Farmers testify that hull-less barley has been declining in frequency, an observation that is substantiated when early records are compared with more recent ones. Some botanical types of the hull-less group identified by Orlov (1929) and Ciferri (1944) no longer occur in the areas where they were once found, demonstrating that genetic erosion has taken place. Hull-less types still occur in Shewa, Gonder, and Tigray. The farmers claimed that their frequency has diminished, and they are being replaced by covered types, which they regarded as hardier and higher yielding. Some hull-less and some partially hulled types owe their existence to women who cultivate them in small plots around living quarters with loving care. While men generally consider the hull-less types more demanding in the field, low yielding, and short lasting, women value them highly as they are less labor intensive to prepare. Recent studies on the nutritive value of hull-less barley, with respect to proteins, fats, minerals, dietary fiber, and energy content (Heen et al. 1991) support the traditional practice of cultivating this barley type for human consumption and their conservation is a critical matter in Ethiopia.

Partially hulled barley. This constitutes a diverse group of two-row barley with lax and dense forms, for which the husk is easily removed upon heating. Partially hulled types occur in many regions, but most frequently in the highlands of Shewa, Gamo Gofa, Gonder, and Bale. In one locality in Shewa (Jibat), 6 (19%) of 31 distinct morphotypes featured partially hulled caryopsis (Asfaw 1990). Pure stands of hull-less barley are observed with higher frequency and wider distribution than the other main types. Partially hulled grains are consumed primarily as roasted grains, which are easy to prepare and simple to serve, requiring light roasting and pounding (dehulling). This is a characteristic reflected in its popular name, *senefgebs,* which means "the lazy person's barley" in the Amharic language. Though grains of other barley types can also be roasted, partially hulled types are of high roasting quality, attributed to the well-developed big and plump grains produced by the central florets of lax spikes.

The popularity of roasted barley among Ethiopians of all ages and the ease with which it can be served at social gatherings, as a "waiting food," and for daily and household consumption contributes to the continued cultivation of partially hulled types. It is widely sold and consumed at bus stops, in drinking houses, and at various social gatherings such as condolence sessions, religious and traditional gatherings in churches, villages, and individual residences. Monks, nuns, and hermits in monasteries and isolated churches live largely on roasted grains of barley, supplemented by wild fruits. Roasted barley is a good traveling food as it may be stored for long periods of time. Usually, roasted barley grains are served mixed with limited quantities of roasted safflower, chickpeas, peas, groundnuts, or roasted and crushed niger-seed balls, all of which improve both the taste and nutritive value. Recently, roasted barley grains have become more widely available in pastry shops and incipient export activity is already underway. Such market value will continue to favor the conservation of this group through cultivation.

The diversity of the barley cultivated in Ethiopia has been affirmed by analysis of its morphology (Asfaw 1988, and literature cited therein), biochemical composition (Bekele 1983; Asfaw 1989c; Demissie 1996), presence of disease-resistant genes (Negassa 1985; Hoyt 1988), and protein and lysine content (Jonassen and Munck 1981; Munck 1992b). The hordein polypeptide pattern is a very useful tool for assessing the range of diversity (Figure 4.6). Different morphotypes vary in their hordein pattern and, in some cases, hordein polymorphism is seen within a single morphological type. More than 40 major hordein groups have been identified, closely matching the degree of morphological variation (Asfaw 1989c).

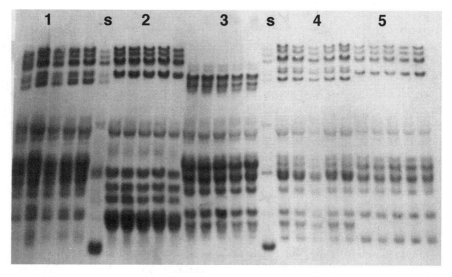

Figure 4.6   Hordein polypeptide pattern in Ethiopian barley. (Each set of 5 columns (1–5) shows patterns of grains from different morphotypes and at positions between 1 and 2 and 3 and 4 are the patterns of the standard cultivar used for comparison and calibration) (Asfaw 1989c).

Factors behind barley diversification in Ethiopia

The great variation and endemicity in barley forms has been interpreted in different ways. N. I. Vavilov initially considered Ethiopia as the center of origin for barley and later on as a secondary diversification center for the crop. The main reason for this reversal of opinion was the fact that the existence of the wild progenitor in Ethiopia has never been confirmed. In some cases, subsequent research has supported Vavilov's determination of Ethiopia as a secondary center (Takahashi 1955; Huffnagel 1961). Other studies favor of the earlier view, particularly with reference to the diversity and endemicity of forms coupled with the frequency of resistant genes for various categories of diseases. The unique endemics such as the *deficiens* and *labile* (irregular) types and the abundance of forms with features that are generally considered primitive in barley (discussed above) have been cited as evidence for the origin of barley in Ethiopia.

While some researchers ascribe at least some of these features to early introduction, others still consider them additional evidence for the origin of barley in Ethiopia. In his study on the biology of cereal landrace populations, Bekele (1985) discusses polymorphism and the balance of forces maintaining overall barley polymorphism in Ethiopia — mutation and selection, selection and migration, the heterogeneous environment, neutral polymorphism, frequency dependent selection, and transient polymorphism as forces. Asfaw (1989a) notes that a combination of agroclimatic and biological processes together with anthropogenic factors is behind the diversification of barley forms in the Ethiopian biophysical and sociocultural environment. The biological processes of natural selection are combined with barley's predominant selfing and limited outcrossing breeding characteristics.

The domestication process, agricultural systems, the agglomeration of different types within single fields and the deliberate selection of lines exercised by farmers have all contributed to the process of fixing characters and maintaining existence within the gene pool. While the selective pressures favor the preservation of many botanical forms, they simultaneously select against other types that consequently became less and less frequent, and even perhaps "extinct" from cultivation at present. Types reported as common in some regions and localities during the Vavilovian expedition (e.g., many naked forms including smooth-awned types) are absent or rare in those areas at the present time.

The net effect of the overall process, however, is the preservation of more types within the agricultural system. In a recent study, Feyissa (1995) supports the view that farmer selection is inversely related to genetic erosion and directly related to conservation. When farmers select, they do not select for a single character. They select for many characters, actually for combinations of characters in a given material, and these characters are directly related to adaptability, yield, nutritional values, and others of utilitarian importance. Though farmers' types often display morphological uniformity, they are not genetically uniform, in sharp contrast with breeders' types. The process helps to actually conserve those desirable characters through cultivation. This is the reason for usually finding many different types in the same field. Barley is famous for such wide phenotypic diversity, which also signifies biochemical and genetic diversity.

Conservation through cultivation is the very subtle strategy of traditional farmers yet to be understood and appreciated by the modern scientific sector. In fact, since different farmers and farming communities select for different sets of characters, the overall diversity sampled for maintenance is very high, as the number of combinations and permutations is tremendously high. Hence, selection as practiced by the traditional barley farmer in Ethiopia does not result in genetic erosion but conserves the full range of the diversity in a dynamic state. This is the ideal breeding strategy for smallholder farmers and those who use the produce largely for consumptive purposes. The nutitional balance attributed to such a genetically broad-based material is the hidden merit that farmers are beginning to realize.

Sociocultural aspects

Study of traditional sayings, lines in poems, beliefs, value systems, and whims shows the significance of barley in the life of Ethiopians to the extent that barley is locally referred to as "the king of grains." The various traditional ethos on barley are also sources of valuable indigenous knowledge because they refer to attributes such as growing habits, seed quality, food quality, brewing quality, character transmission, maturity, and yield (Asfaw 1996). Some of the vernacular names and sayings provide distinguishing attributes for particular morphotypes or landraces referring mostly to distinct botanical features. The naming system is organized under a hierarchical system that is often very descriptive (Asfaw 1990). Traditionally, at the highest level barley is grouped into three tiers: hulled, hull-less, and partially hulled. For example, a traditional classification/nomenclatural series apparent within the partially hulled category designated as *senefgebs* recognizes one form called *senefnetchgebs-balekaport*. Three main botanical features are palpable in this name and they are in a hierarchical order: first, the barley is of the partially naked type; second, it is straw yellow; and third, it has broad outer glumes that cover the grain as an overcoat (Asfaw 1996).

Indigenous knowledge and modern science should be integrated to compile a modern database on Ethiopian barley. It is important that the indigenous knowledge on barley is collected and analyzed through ethnobotanical studies in order to enhance the conservation and use of local landraces of barley both for cultivation and breeding work. Gene banks should make ethnobotanical information part of both their routine collecting formats and their database systems. This strategy will optimize the use of the wealth found within the barley of Ethiopia.

Barley improvement in Ethiopia

Traditional breeding systems

In Ethiopia, barley is cultivated under a small-scale, mixed farming system in the traditional way, which allows for the operation of the natural breeding system. The predominance of inbreeding with some outbreeding is facilitated in the traditional barley cultivation system when different genetic types are grown as mixtures. Changes in genotypic and phenotypic characters under such a system occur gradually, allowing for retention of the wild-type character as well as some of the rare variants in the population for an extended period. The natural breeding system continues, minimally steered by traditional cultivation, seed selection, breeding, harvesting and storage methods. Farmer selection and breeding is a rather subtle process and it can be seen in farmers' maintenance of pure stands, harvesting of better sections of the field for seed, and selecting from the core section of the threshed lot wherein the best seed is found for seed.

Farmers have developed means for correcting deterioration of their barley germplasm. If they believe that the seed is no longer good, they obtain better seed material from known stocks through exchange with relatives or friends. The high quality seeds are usually brought in from the agroecological ranges where the diversity is high and the growing conditions are stable, to provide healthy and more developed grains within the range of genetic variability. Farmers know that the highlands are reservoirs of high quality barley seed as the performance of the crop is consistently better than in other regions of Ethiopia. In the highlands, environmental conditions allow for the expression of a wide array of genes and, therefore, a wide diversity of barley types. Farmers residing in lower altitudes, where growing conditions are more erratic, occasionally revitalize their barley with better quality by exchanging seeds from the highlands. The highlanders usually maintain their original seed stock unless they discover some deterioration in the germplasm in which case they seek better materials from friends or relatives in the village. Farmers who have excess seed material market their seed at the onset of the sowing season when prices are highest.

As a result of seed selection and exchange, a landrace is generally defined as a cultivated (domesticated) population that is genetically heterogeneous and has, over many generations, become adapted to the local environment and cultural conditions under which it is grown. This notation abates the active involvement of farmers in the evolution of landraces, giving the bulk of the credit to the land. The reality is that landraces are produced by farmers and farming communities through traditional breeding practices and should be called farmers' varieties to give due credit to farmers' innovative skills in selecting and cultivating special types. Farmers' varieties represent that special biodiversity found at the interface between absolute wild plant species and the fully domesticated biota under intensive human manipulation. Farmers have mixed and selected, as the case may be, to nurture the landraces that they have maintained.

Modern barley breeding in Ethiopia

Conventional barley breeding began in Ethiopia in 1955, at the College of Agriculture and Mechanical Arts, now the Agricultural University of Alemaya. The coordination of barley research was taken over by the Institute of Agricultural Research, which has implemented breeding and improvement programs at different research stations throughout the country. National barley research has focused primarily on breeding using exotic lines, such as the adaptational breeding of malting barley lines. Trials on exotic food barley lines have met with limited success. The local barley types have not received sizable attention from national research initiatives; rather, local types have been studied largely by foreigners and some staff of the Addis Ababa University. It is reported, however, that over 80% of the barley produced in Ethiopia is derived from farmers' varieties (Alemayehu and Gebre 1987). Increased dissatisfaction with exotic barley material has recently redirected the attention of national barley breeders to local material and research is now underway with the hope of developing some elite material, as with durum wheat (Bechere and Tesemma 1997). Landrace improvement has long been recommended as a strategy for crop improvement (e.g., Qualset 1981) but modern agriculture has lagged behind in this regard. The new direction taken by durum wheat improvement, the current awareness of barley breeders, and the on-farm activities taking root at the Institute of Agricultural Research signal progress in this regard.

Overall, these efforts contribute to the *in situ* conservation of barley landraces in a dynamic process where the modern and traditional systems are dovetailed. The impact will be profound as these efforts will also help to restore traditional farming systems and the associated practices such as crop rotation, intercropping, and seed exchange systems. Additionally, the barley breeding strategy will be reshaped when the participatory breeding program which includes farmers' criteria comes into full swing. The disappearance of traditional landraces has been one of the reasons for erosion of traditional knowledge on farming practices. In some parts where the partially naked barley is no longer cultivated, families are forced to prepare roasted barley food from poor quality grain through an intensive dehulling and pounding process. It is reported that younger generation farmers have no knowledge about some agricultural operations such as rotation cycles and seed rates of landraces since what they know is related to the modern package system (Bechere and Tesemma 1997). The basis for giving due consideration to indigenous barley material in future research and improvement efforts — both in formal breeding programs and in mass selections — is to develop modern cultivars and elite materials and enhance the barley gene pool in the country. The search for high yielding lines, be they landrace enhancement or developing modern cultivars, should continue in appropriate sites and localities in a holistic manner to simultaneously and effectively address conservation and food security issues.

Barley conservation in Ethiopia

Ex situ *conservation*

Ex situ conservation involves the management of living organisms outside their natural habitat. Although the typical example of *ex situ* conservation for crop varieties is that of preservation in modern gene banks, some of the practices involving seed storage and exchange by traditional societies can be interpreted as incipient forms of *ex situ* conservation. Farming communities have a network of collective and individual seed maintenance systems. *Ex situ* conservation in the modern era includes activities of gene banks, botanical gardens, field gene banks, and other systems where germplasm is regularly collected, evaluated, and maintained.

As discussed above, Ethiopian barley germplasm was first collected around the turn of the present century by foreign expeditions including those of Chiovenda and Ciferri (from Italy) (Chiovenda 1912; Ciferri 1940, 1944) and Vavilov (from Russia) (Orlov 1929; Vavilov 1951). Further collections have been made by American and British collecting missions (U.S. Operation Mission to Ethiopia 1954; Huffnagel 1961). Early collections of Ethiopian barley are still maintained in gene banks in the U.S., Germany, Russia, and Italy among others (Orlov 1929; Ciferri 1940, 1944; Negassa 1985). In 1976, the Ethiopian national gene bank was established, with barley germplasm collection and conservation as one of its top priorities. The gene bank has also incorporated among its holdings some accessions of repatriated material, through international and bilateral cooperation. The total current holdings of the gene bank include nearly 14,000 accessions of Ethiopian barley (Demissie 1996).

It has recently come to the attention of those involved with crop conservation that *ex situ* conservation must be complemented with *in situ* methods in order to conserve the genetic material with the dynamic evolutionary processes and the valuable cultural practices and knowledge systems. Furthermore, the need to collect indigenous knowledge along with germplasm of indigenous crops for better utilization and understanding is being increasingly emphasized (see Guarino 1995).

In situ *conservation*

In situ conservation is a strategy of managing living organisms in their natural state and within the natural habitat. It is a system for maintaining genetic resources with due consideration of the natural ecological and agroecological systems to ensure continuation of co-evolutionary processes. In cultivated plants, *in situ* conservation is best referred to as on-farm crop conservation. On-farm conservation involves cultivation of local crop varieties by farmers with support and monitoring from the modern formal sector. Although barley landraces continue to be conserved on-farm through traditional means, growing pressure from the modern agricultural sector, land degradation, and associated environmental problems, famine, and cultural dilution have escalated the state of genetic erosion. Consequently, farmers are forced to abandon their traditional landraces. The traditional system would need to be maintained and further developed to be rewarding for communities; for this, a modern approach is needed. The on-farm conservation scheme allows for the cultivation of the crops in heterogeneous populations, in heterogeneous agroecosystems, and with varied cultural practices. This will allow for the co-evolution of crops with diseases and pests. The value of this conservation strategy is that it carries a component of security in times of diseases or pest outbreaks, as some lines are likely to be resistant to such outbreaks.

Traditional barley conservation in Ethiopia is, in essence, an *in situ* system where the germplasm is maintained by being planted continuously from season to season in the locality of its evolution. Traditional local off-farm conservation is closely associated and strongly linked with the traditional on-farm conservation, through a farmer information network. The scheme is a collective action in which farmers and farming communities maintain the diversity of barley by planting the range of landraces in appropriate localities and micro-agrohabitats within the community so that the germplasm can be located somewhere within the bounds of that community, or sometimes in neighboring communities.

Both the agricultural and social systems contribute to the success of on-farm crop conservation efforts: the former, in terms of existing environment and cultural practices; and the latter, in reference to local seed exchange and farmer selection as well as the indigenous knowledge base in support of the process. In recent years, appreciation for the special value of on-farm crop conservation has grown considerably. In particular, the realization that evolutionary processes are arrested by *ex situ* conservation has drawn increased attention to *in situ* conservation. The signing of the Convention on Biological Diversity in 1994 and global and national policies have highlighted the importance of *in situ* conservation and pledged to support such efforts. In modern *in situ* crop genetic resources conservation the stakeholders include farmers, gene banks, researchers, and scientists. Ethiopia provides a unique set of conditions, including the accumulation of diversity, ecogeographic position, agroecologic diversity, and traditional practices of farming and crop management to make it an ideal place for modern *in situ* conservation of many crops, including barley. On-farm barley conservation provides a unique opportunity for supplementing traditional practices with a modern approach and for developing the scientific parameters of the on-farm method.

On-farm conservation and its relevance to Ethiopian barley

The history of farming is also the history of crop genetic resource management, particularly in the case of barley in Ethiopia. Genetic variations of global significance have originated at the Ethiopian local farm and rural community level, as can be illustrated with the famous examples of the barley yellow dwarf virus resistance (Hoyt 1988) and the high-lysine, high-protein barley gene (Jonassen and Munck 1981; Munck 1992b). Under natural conditions, genes exist, mutate, and increase or decrease in response to dynamic interactions with the soil, climatic factors, diseases, pests, competitors, and human selection. These dynamic interactions extend over the entire agricultural history of barley and over the whole area of its distribution. The primary conservational value of the on-farm strategy is that it fosters this dynamic process.

Crop conservation in Ethiopia has a long history and the system is on-farm conservation (e.g., Worede 1992). Farmers have been the active actors in this process. Pressure from different spheres has in recent years undermined farmers' practices so that the status of crop biodiversity is heading toward erosion and deterioration. Considering the longstanding precedence of informal crop genetic resources management, the integration of local farmers into the international conservation process through joint ventures in on-farm landrace conservation and enhancement schemes would help to enhance agrobiodiversity. This observation has been realized by the scientific sector so that farmers, scientists, and extension workers have been engaged in a program of dynamic on-farm crop genetic resources conservation since 1988 (Worede 1992).

The Ethiopian on-farm landrace conservation and enhancement program is a participatory project (cited above) involving the gene bank, breeders, scientists, and farmers that started in project sites in Shewa and Tigray and is now operating in six sites in parts of Shewa, Tigray, Kefa, Welo, and Bale. Barley is included in project sites located in Shewa, Tigray, and Bale. The project aims to support and encourage farming communities to maintain barley landraces with the associated indigenous knowledge. Farmer conservators are main targets for obtaining the traditionally cultivated landraces and their knowledge of plant characteristics. Farmers are encouraged and supported to obtain such barley landraces and conserve them as they were maintained in the past, including practices such as seed selection and exchange systems.

Indigenous knowledge held by farmers and communities is studied and documented through ethnobotanical surveys and studies. The conservation program focuses on the association of barley with other crops, interaction among crops and varieties within a crop, cultural practices, and factors that safeguard the integrity of the various interactions. Seed maintenance and exchange systems are studied and augmented by the establishment of low-cost community seed banks that operate mainly through the traditional system. Experimental plots are maintained for farmers to evaluate the germplasm for yield, diseases and pests, and other parameters. The operation of this system in Tigray described by Berg (1992) can be taken as an illustrative example. Traditionally, barley has been conserved in Ethiopia by farmers and farming communities, largely on-farm through continuous cultivation individually or within the community and in grain stores, pots, and bottle gourds.

Continuous cultivation is actually a factor in the evolution of new recombinants. Farmers generally keep some seed material for planting, by replanting it immediately or by securely storing it until the next growing season. This practice is supplemented by the community's invisible seed exchange network that ensures a given landrace is kept secure somewhere within the community. Additionally, there is also a local communication system which functions through daily conversation or social gatherings to trace and locate the whereabouts of desired types. The barley farmer may pass along information about the qualities of his barley seed and, hence, indirectly advertise it or express wishes to exchange it with high grade seed of another variety or another crop. This is similar to a farmer-based seed certification mechanism. Taken collectively, the system constitutes a traditional *in situ* conservation strategy, combined with a traditional *ex situ* strategy, which is fully supported by a local information network. Since the seed exchange system of the traditional non-formal sector usually ranges over a short distance, it can be considered part of the *in situ* system and that of the on-farm package. It follows that the best way to implement crop germplasm conservation programs today is to systematically combine *in situ* and *ex situ* conservation strategies. In gene banks, time is frozen at the time of collection and space is squeezed to the small area required to regenerate collected material, displaced from pests and diseases occurring in its natural environment. Hence, regeneration itself aggravates the level of genetic erosion. The new conservation model links farmers and farming communities with formal germplasm conservationists, such that they can learn from and assist each other. In the same manner, barley breeders will be linked with this system to complete the loop of farmer-gene bank-breeder partnership (Figure 4.7). The process will allow for reciprocal exchange of information and germplasm between the informal peasant sector and the formal sector for a mutual benefit.

Appraisal of in situ *conservation of barley*

In situ conservation cannot be viewed independently from production. Traditional Ethiopian barley farmers undertake the production and conservation of landraces simultaneously. This traditional system is of particular merit for barley because of the wide use of diversity and distribution of germplasm among farmers with individual and collective responsibilities. Seed systems of the modern era have interfered with the traditional system by unlinking the seed maintenance system from the production system. The general trend over the past few decades has promoted modern cultivars and crops other than barley, which has led to the gradual erosion of barley's genetic base. This was further aggravated by land degradation, drought, famine, and overall deterioration of environmental vigor and integrity. Consequently, *in situ* conservation of the present time cannot rely solely on the traditional system. Scientists, research institutes, and gene banks should play a supporting role to facilitate a modern *in situ* conservation strategy in the context of the existing on-farm system (Geneflow 1992). Within a framework of conservation, intervention is necessary to improve the quality of the material cultivated in terms of yield, nutritional content, disease resistance, and other attributes with attention to farmer criteria. In this way, the traditional and the modern systems support each other in embracing on-farm conservation strategies. A set of principles would include:

- Grassroots involvement to ensure preservation of the high level of diversity in Ethiopian barley;
- Promotion of small-scale farming, which is based on environmental heterogeneity and in turn favors barley diversity through new combinations of genotypes and alleles;

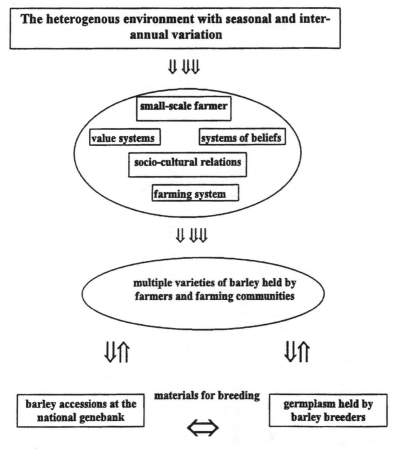

Figure 4.7 Actors and linkages in modern on-farm barley conservation in Ethiopia [Adapted for barley from the conceptual model for genetic conservation presented by van Oosterhout (1994; Figure 5)].

- An intermarriage between traditional knowledge and modern science;
- Integration of farmers' indigenous selection practices and character recognition skills with formal breeding;
- Integration of farmers' breeding strategies and selection criteria with those of the formal sector; and
- Complementary roles for *ex situ* and *in situ* strategies.

The on-farm barley conservation work in operation is a component of the *Dynamic Farmer-Based Approach to the Conservation of Ethiopia's Plant Genetic Resources Project*. Alluded to earlier, this is an innovative approach to a modern integrated *in situ* and *ex situ* conservation, based on partnership between farmers and the national gene bank with support from barley breeders and other scientists. Drawing on the principles outlined above, the project first identifies suitable areas for barley on-farm conservation based on criteria that include the extent of genetic erosion, the history of barley cultivation in the general area, and current levels of diversity.

Knowledgeable barley farmers (conservationists) are carefully identified on the basis of what landraces they have been conserving which are either lost or on the verge of disappearance from the area and their knowledge about such landraces and general crop husbandry in the area. Such farmer partners are briefed about the project and formally invited to become partners in barley conservation. Farmer partners are systematically selected to embrace and will-fully encourage participation of men and women, and older and younger members of the community. Participating farmers identify potential landraces for conservation programs and offer indigenous knowledge of barley and the landraces grown in their particular locality. The project assists farmers so that they can conserve a number of barley landraces through mutual agreements and benefits, including market and non-market incentives. For example, farm-ers are compensated for lower gains in crop return if they happen to harvest less than what a farmer who planted modern cultivars gets. In addition, gene bank materials are made available to farmers for restoration if they wish to take them. Farmer partners are encouraged to continue cultivating landraces according to traditional farming practices, such as crop rotation, organic farm-ing, and seed selection, storage, and exchange systems. The conservation model opts, therefore, to conserve the crop diversity with the valuable biolog-ical processes and traditional practices.

This barley conservation scheme focuses not only on the local varieties of ancestral crop populations, but also on the human knowledge and behav-ioral practices that have shaped this diversity for generations. In the second phase, farmers who have for one reason or another lost their traditional barley landraces, but are now interested in regaining some of them, will be incorporated into the project and assisted in conserving reintroduced landraces. Project assistance includes covering the cost of seeds that are purchased from farmers identified in the first phase and some technical advice. The project will also set up small-scale, low-cost community seed banks in each locality to be managed and used by the farmers, an activity already underway in the Tigray region and for which preparations are under-way in others. In another related effort called the landrace restoration effort, gene bank accessions of landraces collected in the locality some years back are now grown in project sites within small demonstration plots for farmers to see the different types that were at one time cultivated in the area. If farmers show interest in some of these activities, the seed can be multiplied and distributed accordingly.

Indigenous knowledge on these materials will also be collected as farm-ers often recall the types that used to grow in the area; the indigenous knowledge of the landraces survives with the people, even if the landrace itself no longer does. In association with the on-farm conservation drive, Ethiopia is pursuing what is termed the landrace enhancement scheme, which opts to improve promising landraces using farmers' selection criteria

mainly through mixing morphotypes of desirable qualities. The plant structure, yield, disease resistance, and other features of barley can be targeted
for improvement to develop competitive production levels. In this respect,
the barley on-farm conservation effort is following the example of the durum
wheat landrace enhancement scheme, under operation for many years now,
where mixtures of high yielding combinations are reported to have been
already released (Bechere and Tesemma 1997) to farmers in collaboration
with a national nongovernmental organization — Seeds of Survival. Farmer
partners are encouraged to practice the on-farm conservation strategy with
creativity and intuition. Gene banks and researchers will periodically monitor the level of genetic diversity to observe changes in time and space. The
International Plant Genetic Resources Institute (IPGRI) has already drafted
a project proposal to study the scientific basis of on-farm landrace conservation to support the ongoing scheme.

The barley on-farm conservation program is being implemented mainly
in the rugged high altitude regions, considered marginal for most other crops
and unsuitable for high yielding modern barley cultivars. Thus, on-farm
conservation and the continued use of landraces will not interfere with large-
scale production of the crop under high input agriculture, but instead will
seek to enhance agricultural systems in marginal areas. Barley landraces and
high yielding cultivars can co-exist in Ethiopia's agricultural system, thereby
contributing to food security from two angles: product diversification and
high production. An agricultural system that conserves the indigenous landraces in some areas of the country and uses high yielding modern cultivars
in other areas would help to maintain high diversity in that crop while also
increasing production and productivity (see Asfaw 1989a:24). Conservation
of barley on-farm can be implemented step-by-step in parts of Ethiopia
where the genetic resources of the crop are still abundant. Restoration programs can also be implemented in areas where barley was at one time highly
diverse, but has eroded in recent years. Although the genetic diversity of
barley has been drastically reduced across such areas, the range of landraces
may still occur with few farmers within the locality.

The on-farm strategy allows for a two-way flow of barley germplasm
between farmers and gene banks (Geneflow 1992). Researchers and scientists
associated with barley research and breeding can also be linked to this system
for mutual benefits. The factors that have contributed to the diversification
of barley landraces in Ethiopia range from the natural to the sociocultural;
landrace conservation would require due consideration of these same factors
(Asfaw 1989a). Studies of other crops confirm this observation, as with maize
(Bellon 1996) and sorghum (van Oosterhout 1994). In the case of Ethiopia,
numerous processes and systems have been linked and further linkages
should be introduced to fully address the dynamics of barley conservation
within the country (Figure 4.7).

The on-farm conservation process has the special merits of preserving the
genetic diversity of the crop while it is in dynamic adaptation with the agroecosystems and in harmony with the traditional practices and knowledge. It

is also open to the influx of modern scientific knowledge and breeding mate-rials from gene banks, scientists, and researchers. The modern sector will benefit from this partnership by having access to the indigenous landraces and the knowledge base. The traditional system will be able to reap the fruits of modern science without being disadvantaged by it. Other groups including professional societies and non-governmental organizations can promote tra-ditional landrace conservation schemes by raising and distributing seeds of those of interest for conservation. Hence, there will be an active interplay between the traditional and the modern systems. The synergistic effect obtained from the combined input of all the stakeholders and the possibility for operation of all processes will introduce into the system a unique set of advantages.

References

Åberg, E. 1940. The taxonomy and phylogeny of *Hordeum* L. Sect. Cerealia Ands. With special reference to Tibetan barleys, *Sumbolae Botanicae Uppsaliensis* IV:2.

Alemayehu, F. and H. Gebre. 1987. Barley breeding in Ethiopia, *Rachis: Barley and Wheat Newsletter* 6:13–15.

Altieri, M.A. and L.C. Merrick. 1987. *In situ* conservation of crop genetic resources through maintenance of traditional farming systems, *Economic Botany* 41:86–96.

Altieri, M.A. and C. Montecinos. 1993. Conserving crop genetic resources in Latin America. In *Perspectives on Biodiversity: Case Studies of Genetic Resource Conserva-tion and Development*, S.S. Potter, J.I. Cohen, and D. Janczewski (eds.). Washing-ton, D.C.: AAAS Press.

Alvares, F. 1961. *The Prester John of the Indies, Volume 1*. Cambridge: Cambridge University Press.

Anderson, B., Q. Zue, R. Newman, and W. Newman. 1991. Serum lipid concentrations of chickens fed diets with flour or red dog from different types of glacier barley, *Barley Genetics* VI: 461–465.

Asfaw, Z. 1988. Variation in the morphology of the spike within Ethiopian barley, *Hordeum vulgare* L. (Poaceae), *Acta Agric. Scand.* 38:277–288.

Asfaw, Z. 1989a. The barley of Ethiopia: a focus on the infraspecific taxa, *Acta Uni-versitatis Uppsaliensis: Comprehensive Summaries of Uppsala*. Dissertation from the Faculty of Science, Uppsala.

Asfaw, Z. 1989b. Relationships between spike morphology, hordeins and altitude within Ethiopian barley, *Hereditas* 110:203–209.

Asfaw, Z. 1989c. Variation in hordein polypeptide pattern within Ethiopian barley, *Hordeum vulgare* L. (Poaceae), *Hereditas* 110:185–191.

Asfaw, Z. 1990. An ethnobotanical study of barley in the central highlands of Ethiopia, *Biol. Zent. Bl.* 9:51–62.

Asfaw, Z. 1991. Hybrids between *Spontaneum* and Ethiopian barley, *Barley Genetics* VI (1):56–61.

Asfaw, Z. 1996. Barley in Ethiopia: the link between botany and tradition. In *Barley Research in Ethiopia: Past Work and Future Prospects*, H. Gebre and J. van Leur (eds.). Proceedings of the first barley research review workshop, 16-19 October 1993. Addis Ababa. IAR/ICARDA.

Asfaw, Z. and R. von Bothmer. 1990. Hybridization between landrace varieties of Ethiopian barley (*Hordeum vulgare* ssp. *vulgare*) and the progenitor of barley (*Hordeum vulgare* ssp. *spontaneum*), *Hereditas* 112:57–64.

Bechere, E. and T. Tesemma. 1997. Enhancement of durum wheat landraces in Ethiopia. Workshop paper, Biodiversity Institute, Addis Ababa, Ethiopia.

Bekele, E. 1983. A differential rate of regional distribution of barley flavonoid patterns in Ethiopia, and a view on the centre of origin of barley, *Hereditas* 98:269–280.

Bekele, E. 1985. The biology of cereal landrace populations. 1. problems of gene conservation, plant-breeding selection schemes and sample size requirements, *Hereditas* 103:119–134.

Bell, G.D.H. 1965. The comparative phylogeny of the temperate cereals. In *Essays in Crop Plant Evolution*, Sir J. Hutchinson (ed.). Cambridge: Cambridge University Press.

Bellon, M.R. 1996. The dynamics of crop infraspecific diversity: a conceptual framework at the farmer level, *Economic Botany* 50:26–39.

Berg, T. 1992. Indigenous knowledge and plant breeding in Tigray-Ethiopia, *Forum for Development Studies* 1:13–22.

Bhatty, R.S. 1992. Dietary and nutritional aspects of barley in human foods, *Barley Genetics* VI 2:913–924.

Brandt, S.A. 1984. New perspectives on the origins of food production in Ethiopia, In *From Hunters to Farmers: the causes and consequences of food production in Africa*, J.D. Clark and S.A. Brandt (eds.). Berkeley: University of California Press.

Briggs, D.E. 1978. *Barley.* New York: Chapman & Hall.

Burger, W.C, A.A. Quresffi, and N. Prentice. 1981. Dietary barley as a regulator of cholesterol and fatty acid metabolism in the chicken, rat and swine, *Barley Genetics* 4:644–651.

Chiovenda, E. 1912. Etiopie osservazioii botaniche agrarie ed industriali, *Monografie Raporti Coloniali Typografic Nazionali, Di G.* Rome: Berfero E.C.

Ciferri, R. 1940. Saggio di classificazione degli orzi con speciale reguardo A. quelli Etiopici, *Nuovo Giomale Italiana* 47:423–434. Societa Botanica Italiana.

Ciferri, R. 1944. Osservazioni ecologico agrarie e sistematiche su painte coltivate in Etiopia, 20 C-1(38):179–200.

Clark, J.D. and M.A.J. Williams. 1978. Recent archaeological research in southeastern Ethiopia (1974-1975), *Annales D'Ethiopie* 11:19–44.

Demissie, A. 1996. Morphological and Molecular Marker Diversity in Ethiopian Landrace Barleys: Implications to In-situ and Ex-situ Conservation of Landrace Materials. Ph.D. Dissertation. Agricultural University of Norway.

Demissie, A. and A. Bjornstad. 1996. Phenotypic diversity of Ethiopian barleys in relation to geographical regions, altitudinal range, and agroecological zones: as an aid to germplasm collection and conservation strategy, *Hereditas* 124:17–29.

Demissie, A. and A. Bjornstad. 1997. Geographical, altitudinal and agroecological differentiation of isozymes and hordein genotypes of landraces of barleys from Ethiopia: implications to germplasm conservation, *Genetic Resources and Crop Evolution* 44:43–55.

Ehret, C. 1979. On the antiquity of agriculture in Ethiopia, *The Journal of African History* 20:161–177.

Engels, J.M.M. 1986. A diversity study of Ethiopian barley. In *Plant Genetic Resources of Ethiopia*, J.M.M. Engels, J.G. Hawkes, and M. Worede (eds.). Cambridge: Cambridge University Press.

Feyissa, R. 1995. *In-situ* Conservation of Crops: Ethiopian Model. (Workshop paper).

Gamst, F.C. 1969. *The Qemant: A Pagan-Hebraic Peasantry of Ethiopia.* Homewood, IL: Waveland Press.

Gebre, H. and F.F. Pinto. 1977. Barley production and research in Ethiopia. Unpublished paper.

Geneflow. 1992. Farmers do it all the time. Special UNCED Edition: *Biodiversity and Plant Genetic Resources.*

Giles, B.E. and R. von Bothmer. 1985. The progenitor of barley (*Hordeum vulgare* ssp. *Spontaneum*) — its importance as a gene resource. *Sveriges Utsadesforenings Tidskrift* 95:53–61.

Grillot, G. 1959. La classification des orges cultivees (Hordeum sativum Jenssen) et nouvelles varietes d'orge, *Annales de L'Amelioration des Plantes* IV:445–551.

Guarino, L. 1995. Secondary sources on culture and indigenous knowledge systems. In *Collecting Plant Genetic Diversity,* L. Guarino, V. Ramanatha Rao, and R. Reid. (eds.) United Kingdom: IPGRI, FAO, UNEP, IUCN; CAB International.

Haberland, E. von. 1963. *Galla Sud-Athiopiens.* Stuttgart: Verlag W. Kohlhammer. (German with English summary)

Harlan, J.R. 1968. On the origin of barley. In *Barley: Origin, Botany, Culture, Winter Hardiness, Genetics, Utilization & Pests, USDA Agricultural Handbook* 338:12–34.

Harlan, J.R. 1969. Ethiopia: a centre of diversity, *Economic Botany* 23:309–314.

Heen, A., O.R. Eide, and W. Frolish. 1991. Hull-less barley, agronomic traits and nutritional quality, *Barley Genetics* IV:446–448.

Hoyt, E. 1988. Conserving the wild relatives of crops. Rome and Switzerland: IPGRI, IUCN, and WWF.

Huffnagel, H.P. 1961. *Agriculture in Ethiopia.* Rome: Food and Agriculture Organization.

Index Seminum. 1983. *Hordeum: Academia Scientiarum Republicae Germanicae Democraticae.* Gatersleben: Zentralinstitut für Genetic und Kulturpflanzenforschung.

Jonassen, I.T. and L. Munck. 1981. Biochemistry and genetics of the Sp II albumin in hiproly barley, *Barley Genetics* IV:330–335.

Lakew, B., H. Gebre, and F. Alemayehu. 1996. Barley production and research in Ethiopia. In *Barley Research in Ethiopia: Past Work and Future Prospects,* H. Gebre and J. van leur (eds.). Proceedings of the first barley research review workshop, 16–19 October 1993. Addis Ababa. IAR/ICARDA.

Metcalfe, D.R., A.W. Chicko, J.W. Martens, and A. Tekuaz. 1978. Reaction of Ethiopian barleys to Canadian barley pathogens, *Canadian Journal of Plant Science* 58(3):885–890.

Munck, L. 1992a. Summary — nutritional quality, *Barley Genetics* IV(2):945–952.

Munck, L. 1992b. The contribution of barley to agriculture today and in the future, *Barley Genetics* VI(2):1099–1110.

Negassa, M. 1985. Patterns of phenotypic diversity in an Ethiopian barley collection, and the Arsi-Bale Highlands as a centre of origin of barley, *Hereditas* 102:139–150.

Orlov, A.A. 1929. The barley of Abyssinia and Eritrea, *Bulletin of Applied Botany, Genetics and Plant Breeding* 20:283–345. (Russian with English summary)

Orlov, A.A. and E. Åberg. 1941. The classification of subspecies and varieties of Hordeum sativum Jensen, Extract and translation from A.A. Orlov (1936), Hordeum L., "Barley." In *Flora of Cultivated Plants* 2:1–18.

Porteres, R. 1976. African cereals. In *Origins of African Plant Domestication,* J.R. Harlan, J.M. de Wet, and A.B.L. Stemler (eds.). The Hague: Mouton Publishers.

Purseglove, J.W. 1976. The origin and migration of crops in Tropical Africa. In *Origins of African Plant Domestication,* J.R. Harlan, J.M. de Wet, and A.B.L. Stemler (eds.). The Hague: Mouton Publishers.

Qualset, C.O. 1975. Sampling germplasm in a centre of diversity: Aan example of disease resistance in Ethiopian barley. In *Crop Genetic Resources for Today and Tomorrow,* O.H. Frankel and J.G. Hawkes (eds.). EBP 2. Cambridge: Cambridge University Press.

Qualset, C.O. 1981. Barley mixtures: the continuing search for high-performing combinations, *Barley Genetics IV*(2):130–137.

Smith, L. 1951. Cytology and genetics in barley, *Botanical Review* 17:1–51, 133–202, 285–3 55.

Soleri, D. and S.E. Smith. 1995. Morphological and phenological comparisons of Hopi maize varieties conserved *in situ* and *ex situ, Economic Botany* 49(1):56–77.

Takahashi, R. 1955. The origin and evolution of cultivated barley, *Advanced Genetics* 7:227–266.

U.S. Operation Mission to Ethiopia. 1954. *The Agriculture of Ethiopia,* Vol. I. Unpublished Staff Report. Imperial Ethiopian College of Agriculture and Mechanical Arts. Jimma Agricultural and Technical School.

van Oosterhout, S. 1994. The development of dynamic conservation measures for indigenous crop landraces: limitations imposed by the hegemony of western science. In *Safeguarding the Genetic Basis of Africa's Traditional Crops,* A. Putter (ed.). The Netherlands: CTA/Rome: IPGRI.

Vavilov, N.I. 1951. The origin, variation, immunity and breeding of cultivated plants, *Chronica Botanica* 13:1–366.

von Alkamper, J. 1974. The influence of altitude on yield and quantity in cereals in Ethiopia, *Z. Acker und Pflanzenbau* 140:184–189.

von Bothmer, R. and N. Jacobsen. 1985. Origin, taxonomy and related species, *Barley-Agronomy Monograph* 26:19–125.

von Bothmer, R., N. Jacobsen, and R. Jorgensen. 1981. Phylogeny and taxonomy in the genus *Hordeum, Barley Genetics* IV: 13–21.

von Bothmer, R. and O. Seberg. 1995. Strategies for the collecting of wild species. In *Collecting Plant Genetic Diversity,* L. Guarino, V. Ramanthana Rao, and R. Reid (eds.). Rome: IPGRI/Cambridge: Cambridge University Press.

Wendorf, F.S., R. El Hadidi, N. Close, A.E. Kobsiewicz, M. Wieckowska, H. Issawi, and H. Haas. 1979. Use of barley in the Egyptian Late Paleolithic, *Science* 205:1341–47.

Westphal, E. 1975. Agricultural systems in Ethiopia. *Agricultural Research Report* 826. Wageningen: Center for Agricultural Publication and Documentation.

Worede, M. 1992. Ethiopia: A gene bank working with farmers. In *Growing Diversity: Genetic Resources and Local Food Security,* D. Cooper, R. Vellve, and H. Hobbelink (eds.). London: Intermediate Technology.

chapter five

Traditional management of seed and genetic diversity: what is a landrace?

Dominique Louette

Introduction

Increasing concern about the loss of genetic resources over the past 20 years has led to a heightened concentration on methods for conservation of genetic resources in gene banks (*ex situ* conservation) (Bommer 1991). Conservation of the genetic resources in the agrosystem in which they have evolved *(in situ* conservation) is now being more widely considered as complementary to *ex situ* strategies for conserving genetic diversity (Altieri and Merrick 1987; Cohen et al. 1991; Cooper et al. 1992; FAO 1989; Keystone Centre 1991; Merrick 1990; Montecinos and Altieri 1991; Oldfield and Alcorn 1987). *In situ*, or on-farm, conservation has been proposed essentially for wild relatives of cultivated plants or for plants with recalcitrant seeds. When considered for other cultivated species, this alternative (on-farm conservation) continues to be highly polemic, considered unfeasible from a socioeconomic perspective. The model also raises numerous questions about how policies aimed at fostering economic development relate to those designed to conserve plant genetic resources and whether conservation can coexist with the integration of communities into commercial markets (Cohen et al. 1991; Cooper et al. 1992; Montecinos and Altieri 1991).

Discussions on *in situ* genetic resources conservation generally consider the "biological reserve model" proposed by Iltis (1974). This model is based on the belief that the best means for *in situ* preservation of the diversity found in genetic material is to "freeze" the genetic landscape by isolating it in space and time, maintaining intact the technical, social, and cultural context in which it occurs (Iltis 1974; Benz 1988). Cultivation of local varieties would be encouraged and introduction of foreign cultivars and of new techniques would be discouraged. In this on-farm conservation model, local varieties or landraces are identified as the conservation units. A local variety is well defined in space and as the result of local management. It is also genetically defined as there is concern about geneflow or contamination from other varieties in the case of open-pollinated plants. This chapter adopts a different approach to on-farm conservation. The dynamic nature of agricultural systems precludes "freezing" local varieties into a static system, since local varieties exist as part of a dynamic system that extends beyond a single place.

In the indigenous community of Cuzalapa in western Mexico (within the region of origin for maize), traditional maize variety management is not conducted in accordance with the preconceptions of freezing genetic landscapes or focusing on localness. This study examines the structure of genetic diversity in maize and analyzes the effect of farmers' seed management strategies on this structure. Its objective is to determine what farmers conserve of the varieties they cultivate and to specify the mechanisms responsible for the structure and dynamic of diversity in traditional agroecosystems. Two specific questions are examined in this chapter. First, to what extent can the genetic diversity in the maize varieties of Cuzalapa be attributed to the management of materials of strictly local origin? Second, how well defined, genetically, is a local variety of an open pollinated plant?

Data on seed sources illustrate the important role played by seed acquired from other farmers in and outside of the region relative to seed that local farmers obtain from their own harvests. Analyses of phenotypic and phenological characteristics combined with data on the origin of seed demonstrate the effect of introduced varieties on the diversity of maize cultivated in the Cuzalapa community. The amount of seeds used to reproduce the variety, the management of those seeds in space and time, and the traditional selection of seed call into question the genetic definition of a landrace.

The Valley of Cuzalapa and the Sierra de Manantlán Biosphere Reserve

The indigenous community of Cuzalapa is located in a valley in the southern section of the buffer zone of the Sierra de Manantlán Biosphere Reserve (SMBR), in the municipality of Cuautitlán, in the state of Jalisco, on the Pacific Coast of Mexico (Figure 5.1). As the Biosphere Reserve is situated on the Pacific slope of Mexico, most likely one of the zones where the genus *Zea* originated (Benz and Iltis 1992), it is considered an important zone for on-farm conservation of the maize genetic diversity (Jardel 1992). In the reserve and nearby, various species of teosinte, wild relatives of maize (*Zea mays* spp. *parviglumis* Iltis, Doebley; Zea *diploperennis* Iltis, Doebley, Guzmán; and *Zea perennis* Hitchc. Reeves, Mangelsdorf) are found growing alongside

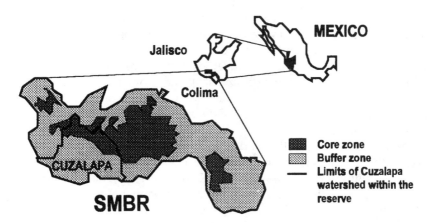

Figure 5.1   Sierra de Manantlán Biosphere Reserve (SMBR) and Cuzalapa Watershed Location within the Reserve.

pre-colonial races of maize such as Tabloncillo and Reventador (Benz 1988; Benz n.d.; Benz et al. 1990; Wellhausen et al. 1952).

   The Cuzalapa watershed covers nearly 24,000 ha (most of which lies within the boundary of the Biosphere Reserve) of mountainous land of extremely irregular topography, ranging from an elevation of 550 to 2660 m. The agricultural zone is located at an elevation of 600 m and is characterized by a hot subhumid climate, with a mean annual temperature of 22°C and mean annual precipitation of 1,500 mm, concentrated from June to October (Martínez et al. 1991). Fields are generally located near rivers on alluvial soils of moderate fertility (Martínez and Sandoval 1993).

   Each year, about 1,000 ha are sown in Cuzalapa, 600 ha of which are irrigated (Martínez and Sandoval 1993). Maize *(Zea mays* spp. *mays)* is the dominant crop in the valley. Nearly half of the survey farmers cultivate maize in association with squash *(Cucurbita* spp.) on an average of 2 ha per farmer during the rainy season, from June to November. Maize is also planted under irrigation in the dry season, which extends from December to May, intercropped with beans *(Phaseolus vulgaris* cv. bayo and bayo berrendo) for the majority of the survey farmers, on an average of 2 to 3 ha per farmer. During this season, a green tomato *(tomatillo* or *Physalis philadelphicum)* grows spontaneously in the fields. Irrigation and intercropping have been common features of agriculture in Cuzalapa since precolonial times (Laitner and Benz 1994). Cultural practices have evolved in Cuzalapa but continue to be relatively traditional when compared to those found outside the Sierra de Manantlán. Farmers generally till arable soils with horse-drawn plows in the rainy season. Tractors are used more frequently during the dry season because at this time, the economic returns to maize production are greater and more reliable, and the irrigated soils contain fewer rocks. Weeds are usually controlled by horse-drawn cultivator before sowing and 1 month after. Sowing, fertilization, and harvesting are always manual operations.

The irrigation system is gravity powered. With these techniques, the survey farmers have obtained mean maize yields of 2.8 tons per hectare (unshelled) during the rainy season and 2.1 tons per hectare (unshelled) in the dry season (under irrigation). Beans are produced exclusively for home consumption. Part of the annual maize crop and almost all of the tomatillo crop of Cuzalapa are sold outside the valley, yet generally the Cuzalapa community is poorly integrated into commercial markets. Extensive cattle raising is now emerging as a commercial activity.

Because of the use of the land by indigenous peoples since pre-colonial periods, the region was officially recognized as a *comunidad indigena* (indigenous community) under the Agrarian Reform of 1950. The valley of Cuzalapa has approximately the same number of inhabitants today (1,500) as it did in 1540 (Laitner and Benz 1994). Now, however, a large proportion of the inhabitants are *mestizos* (of both European and indigenous ancestry). Although it is one of the largest communities of the Biosphere Region, Cuzalapa is also located in one of the most marginalized municipalities of the region, based on quality of housing and level of education (Rosales and Graf 1995). At the time of this study (1989–1991), these localities were all remote from major roads and urban areas. Based on its farming and socioeconomic characteristics, Cuzalapa is representative of many indigenous, poor, and isolated rural areas in Mexico. Cuzalapa is one of the many traditional communities in Mexico which are being drawn slowly into commercial marketing systems while maintaining features of indigenous society.

Varieties and seed lots: flow and diversity

"Seed lot" and "variety" defined

The terms and concepts used in this work are based on farmers' own practices and concepts. In this context, the term "seed lot" refers to the set kernels of a specific type of maize selected by one farmer and sown during one cropping season to reproduce that particular maize type. A "variety" or "cultivar" is defined as the set of farmers' seed lots that bear the same name and are considered to form a homogeneous set. A seed lot, therefore, refers to a physical unit of kernels associated with the farmer who sows it; a variety is associated with a name.

A maize variety is defined as "local" when seed from that variety has been planted in the region for at least one farmer generation (that is, for more than 30 years, or if farmers maintain that "my father used to sow it"). This definition implies that a "local" variety has been cultivated continuously among survey farmers in Cuzalapa for many years. By contrast, an "exotic" variety is characterized either by the recent introduction of its seed lots or by episodic planting in the valley. Exotic varieties may include landraces (farmers' varieties which have not been improved by a formal breeding program) from other regions and commercial improved varieties recently or repeatedly reproduced by farmers using traditional methods.

Seed exchange

Documenting the exchange of seed lots and varieties

To document which maize varieties are cultivated and to record the exchange of seeds and varieties in the community and between the valley of Cuzalapa and other regions, 39 farmers (one fifth of Cuzalapa farmers) were surveyed during six cropping seasons spanning three calendar years (the 1989, 1990, and 1991 rainy and dry seasons). For each farmer and cropping season, data were collected on varieties cultivated and seed source. Cultivars included those grown on the farmer's own fields, on rented fields, and on fields in association with other farmers. Each variety was registered with the name given by the farmer. When the seed introduced from another region shared the same name as a local variety but was not considered, by the farmer growing it, to be the local variety, a second label was noted in brackets (e.g., Negro [Exotic]).

The seed source was classified in three ways: (1) as own seed (seed selected by the farmer from his own harvest); (2) as seed acquired in Cuzalapa (seed obtained in the valley of Cuzalapa from another farmer); and (3) as an introduction (seed acquired outside of the Cuzalapa watershed). The origin of a seed lot is defined independently of the origin of the previous generation of seed. A seed lot is considered "own seed" if the ears from which the kernels were selected were harvested by the farmer in his field in Cuzalapa, even though the seed that produced those ears (i.e., the previous generation of seed) may have originated in another region. The data, therefore, are representative of the extent of seed exchange, but they understate the importance of exotic seed in Cuzalapa.

Regular introduction of exotic varieties

During the six seasons included in the survey, survey farmers grew a total of 26 varieties (Table 5.1). Each farmer grew between one and seven maize varieties during each season and, on average, more than two varieties per season. Most of these cultivars are white-grained dents and are primarily used for making tortillas, the starchy staple of the Mexican diet. Three flinty popcorn varieties (Guino Rosquero, Negro [Guino], and Guino Gordo) were also identified, as well as three purple-grained varieties (Negro, Negro [Exotic], Negro [Guino]) and three yellow-grained varieties (Amarillo Ancho, Amarillo, Amarillo [Tequesquitlán]). The taste of the purple varieties is considered sweeter and the ears of these varieties are generally consumed roasted at the milky stage, while yellow varieties are used essentially as feed for poultry and horses.

Contrary to the general perception of traditional rural societies in relation to cultivated varieties, this community does not function as an isolated area. On the contrary, exotic varieties are regularly introduced for on-farm testing. From the 26 varieties identified, only the cultivars Blanco, Amarillo Ancho, Negro, Tabloncillo, Perla, and Chianquiahuitl are local and all related

Table 5.1   Importance of Varieties Cultivated in Cuzalapa

Varieties	% Maize Area	% Farmers	Grain Color	Cycle Length
6 Local				
Blanco	51%	59%	White	Short
Chianquiahuitl	12%	23%	White	Long
Tabloncillo	5%	6%	White	Short
Perla	0.4%	0.02%	White	Short
Amarillo Ancho	8%	23%	Yellow	Short
Negro	3%	34%	Purple	Short
20 Foreign				
3 most cultivated				
Argentino	5%	10%	White	Long
Enano	3%	12%	White	Long
Amarillo	3%	11%	Yellow	Long
17 minor varieties	<3% per variety	<4% per variety	Mainly white	Mainly long

to the Tabloncillo race (Table 5.1). In other words, only these six varieties had been grown continuously for at least one farmer generation in the valley of Cuzalapa. Only the introduction date of the Chianquiahuitl can be traced to 40 years ago. Four of the six local varieties are cultivated by a large percentage of farmers. Since two of these varieties have white grains (Blanco and Chianquiahuitl), one has yellow grains (Amarillo Ancho), and the fourth has purple grains (Negro), all four varieties provide for the different household uses of maize in Cuzalapa. Although reduced in number, the local varieties cover more than 80% of the area. The two principal white varieties alone occupy an estimated 63% of the area planted to maize.

The remaining 20 of the 26 varieties that Cuzalapa farmers grew during the survey period are classified as exotic. Each exotic variety covered less than 5% of the maize area planted in each season, and most were cultivated by only a few farmers at a time. The composition of this group of varieties changed from season to season. Only three of these varieties (Argentino, Enano, and Amarillo) had been regularly cultivated over the preceding 4 or 5 years by a significant percentage of farmers (10 to 12%). Most had been used for the first time recently or during the survey period and had been planted again once or twice.

The origin of the exotic varieties is often difficult to ascertain. Farmers are able to indicate in which community they acquired a variety, but not its true source. Even the original name of the variety can disappear or take on a different meaning when farmers exchange seed. Based on the information collected, exotic varieties can be classified into three groups: farmers' varieties (landraces) (15); farmers' advanced generations of improved varieties (4); and recent generation of an improved variety (1). The group of exotic varieties is morphologically diverse, including white-, yellow-, and purple-grained materials, and representatives of different races. Most cultivars were introduced from communities of southwestern Jalisco, less than 100 km from Cuzalapa, although the Guino [U.S.] variety cultivated by one farmer originated in the U.S. In general, the data indicate that maize cultivation in Cuzalapa depends notably on local materials but also on a changing and diverse group of exotic varieties introduced through farmer-to-farmer exchanges.

Seed lot exchange

By detailing the geographic origin of each farmers' seed lots, for each variety, in each planting cycle, one can determine and characterize the frequency of seed exchange among farmers and the pattern of variety diffusion. During the study period, the survey farmers sowed 484 seed lots for the total 26 varieties they cultivated, on 442 ha. Many of these seed lots came from other farmers or regions (Figure 5.2). On average, for all cropping seasons, survey farmers selected slightly over half (53%) of their seed lots from their own harvest. About 36% of the seed lots were obtained from another farmer in Cuzalapa, and 11% were introduced from other regions. Calculated in terms of area planted, seed from farmers' own harvests represented 45% of the maize area in the study zone, whereas 40% was planted to seed from other Cuzalapa farmers and 15% was planted to exotic introductions. Seed exchange — whether between farmers inside the valley or with farmers outside the valley — is clearly very important.

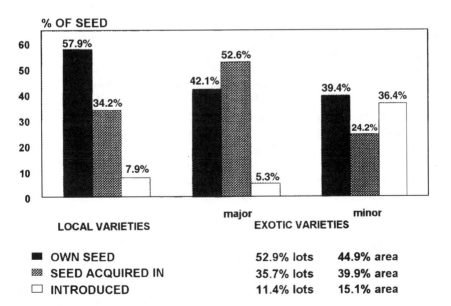

Figure 5.2   Origin of seed planted in Cuzalapa by origin of variety (39 farmers during six growing cycles).

The pattern of varietal diffusion

Both local and exotic varieties were planted from farmers' own seed lots, from seed lots acquired in Cuzalapa, and from introduced seed lots, but in different proportions. Significant differences in origin were associated with the dominance of the variety in terms of planted area (Figure 5.2). Three different groups of varieties were considered: local varieties (6 varieties); most important exotic varieties (3 varieties, Argentino, Amarillo, and Enano); and minor exotic varieties (17 varieties). Seed of the most widely grown varieties, noted in the text as "major varieties" — including the local varieties and the three most important exotic varieties — is less likely to have been obtained from farmers outside of Cuzalapa than seed of the more minor exotic cultivars (7.9% of local varieties and 5.3% of important exotic varieties seed lots were introduced, compared to 36% of minor exotic varieties seed lots). Nevertheless, it is difficult to establish a pattern for the minor exotic varieties because each variety appears to be a special case defined by the time of its introduction and the number of farmers planting it.

Among local varieties, farmers manage the seed for Chianquiahuitl and Negro the most conservatively. More than 70% of the seed for these varieties is selected from farmers' own maize harvests. In fact, farmers plant such a small area to the variety Negro that, on average, seed equivalent to only 27 ears is required per farmer (Louette 1994). This amount of seed, in good condition, is carried over easily from one cycle to the next, and farmers do not need to seek out seed from another farmer. Chianquiahuitl is a variety of unknown origin that is believed to be no longer widely cultivated outside the study zone. Thus, of necessity, farmers in the Cuzalapa Valley must rely on their own stocks.

The case of Blanco, the most important local variety, contrasts with that of Chianquiahuitl. Of all the local varieties, Blanco has the highest proportion of seed obtained from farmers outside the study zone (15%). This result reflects the importance of Blanco in terms of area cultivated in Cuzalapa and nearby regions. Because Blanco is important for household subsistence, an insufficient number of ears suitable for seed may remain at planting time, compelling farmers to search for seed from other farmers in and outside the community.

Both important and minor exotic varieties are also sown from a significant percentage of own seed lots (42.1% and 39.4%, respectively). Farmers reproduce the more important exotic varieties as they do with local varieties. Recently introduced minor varieties are reproduced from farmers' own seed and tested over several seasons. Important and minor exotic varieties can be distinguished by their pattern of diffusion. The percentage of seed brought from other regions is small for the most widely grown exotic varieties (5.3%), while for some of the minor varieties introduced late in the survey period all seed lots were introduced (average 36.4%). Farmers in the valley exchange seed of the important exotic varieties (52.6%) much more frequently than seed of the minor exotic varieties (24.2%). This is explained by the fact that important exotic varieties were introduced some 10 years ago, and because they have demonstrated characteristics of value, their seed is redistributed to other farmers in Cuzalapa. In contrast, survey farmers who did not plant the minor varieties during the study period are presumably not yet convinced of their advantages and do not look for seed.

In summary, there is a moderate level of diffusion of local varieties inside the watershed and little infusion from other regions. Recently introduced exotic varieties are infused from outside the valley. Older exotic varieties that have attained a moderate level of acceptance are also diffused inside the watershed. The pattern of varietal diffusion is therefore linked essentially to the local acceptance of the variety, the time it has been sown in the region, and the availability of seed inside and outside the region. What is important to note is that seed lots introduced from outside the valley can be considered as part of the local varieties. A "local" variety is therefore not constituted by seed lots of exclusively local origin. This finding is important for the concept of a local variety and will be discussed later in this chapter.

Farmer type

The general patterns of maize seed exchange described above nevertheless conceal major differences among survey farmers. Three major farmer groups can be identified. At one extreme are farmers who select seed almost exclusively from their own maize harvests. They sow the same varieties regularly and only modify the proportion of maize area planted to each variety in each cropping season. These farmers are considered suppliers of seed of local cultivars ("they always have seed").

Other farmers use their own seed lots in addition to seed acquired in the community or introduced from other regions, and the proportions of each type of seed vary from season to season depending on each farmers' objectives and constraints. These farmers are generally regarded as suppliers of introduced seed, and some are known in the community for their curiosity about new varieties. At the other end of the spectrum are farmers who have never used their own seed. Throughout the study period, these farmers acquired seed both within and outside of Cuzalapa. This group of farmers includes those who do not have rights to land and cannot plant maize each season and those who farm small areas on which they cannot harvest enough maize for both family consumption and seed. Farmers in this group are obliged to look for seed from other farmers when they want to plant maize.

A relationship exists between the number of varieties (different seed lots) sown by each farmer in each cycle and farmer type or proportion of the farmers' seed stocks originating from his own harvest (correlation coefficient of 0.5). In general, farmers who have more recourse to seed produced by other farmers appear to plant fewer varieties per cycle. For example, the group of farmers who sow more than 90% of their crop with seed from their own harvests planted an average of 2.6 varieties per cycle, while those who used no seed from their own harvests planted an average of only 1.3 varieties per cycle. This finding may reflect either a greater reliance on diverse maize types by more conservative farmers, or it may reflect the fact that searching for seed from other farmers requires more effort and is therefore associated with fewer varieties sown.

Factors explaining seed exchange

Several factors induce farmers to exchange seed. The first is the traditional method of seed storage. Maize (for seed and for food) is stored in bulk in a room of the house. Ears are often attacked by weevils and other insects when the grain is stored for longer than 6 months (from one dry season to another dry season, for example). If a farmer sows a particular variety in only one season per year and has not sown that variety in the previous year, or if the cropping calendar obliges him to plant before harvest, he will search for seed from ears that have been harvested more recently by other farmers. The dry season is better for providing seed because more area is cultivated. Either as a percentage of area planted or as a percentage of total seed stocks, the interchange of seed is more evident at the end of the rainy season. For example, farmers' own maize harvests provide 32 and 57% of the seed for Blanco and Chianquiahuitl grown in the dry season and 69 and 81% of the seed for these varieties during the rainy season.

A second important factor affecting the importance of farmers' seed sources in planting decisions is the socioeconomic status of the household, as represented by farm size, land use rights, and access to the market for renting land. As noted above, many farmers do not cultivate an area large enough to meet their annual food consumption needs, whereas others own no land and must rent a field to cultivate maize. These farm households often consume all of one season's production before planting and are obliged to search for seed each season.

Another factor influencing the seed sources used by farmers is the custom in the Cuzalapa region of producing maize under sharecropping arrangements. Under these arrangements, the partner (or *mediero*) generally supplies labor while the field owner (or *patrón*) supplies the inputs, in particular, maize seed. Generally the *mediero* does not choose which varieties to plant and at harvest time acquires seed from the *patrón*. Seed is also loaned, under the proviso that double the quantity of seed loaned must be returned at harvest. In either case, the farmer obtains maize seed of a variety that another farmer has chosen to grow and that is derived from another farmer's harvest.

Another finding from the survey is that few farmers expressed any particular preference for or allegiance to their own maize as a source of seed. Seed of a given variety selected from their own maize harvest or acquired from other farmers was considered equivalent. In other words, another farmer's method of seed management was not a cause for concern. Furthermore, if a farmer does not grow a particular variety for several successive seasons, this does not signal that the farmer has ceased cultivating it altogether, as long as the seed for that cultivar can still be obtained from other farmers if necessary. Farmers also generally consider that they must change seed regularly to maintain the productivity of the variety ("sow the same maize type but from new seed"). The frequency of seed renewal varies from several cycles to several years. It appears unlikely that any farmer in Cuzalapa sows seed derived from a stock bequeathed directly from his parents.

Finally, farmers appeared to be very curious and open-minded, in general, about testing new cultivars. After visiting a relative or friend, or after harvesting a maize field as a laborer, a farmer often returns with maize ears so that he can test a variety the ear characteristics of which he admires. The introduced seed lots acquired from other farmers are almost never bought as seed. They are gifts from friends or family members living outside the zone or are selected from maize ears bought for consumption.

Phenotypic diversity of varieties

The patterns of maize production and seed management described above are characterized by continual introductions of varieties and, within varieties, considerable exchange of seed among farmers. These findings raise questions about the structure of maize diversity in the Cuzalapa watershed. For example, how can an introduced seed lot be integrated into a local variety? Do exotic varieties compete with local varieties or are they complementary? Analyses of the phenotypic diversity of maize grown in Cuzalapa provide a way to examine some of these questions.

Measuring morphological diversity

The structure of phenotypic diversity was studied both within a variety (among seed lots of a variety) and across varieties (among sets of seed lots bearing different names). Fourteen of the 26 cultivars identified by farmers (all six local varieties and eight exotic varieties) were selected for analysis based on their origin and seed availability. The number of seed lots per cultivar (one to six) varied according to the importance of the cultivar in terms of planted area.

Morphological descriptors were measured in a controlled experiment of maize grown in pure stand in three complete blocks. The experiment was established in a farmer's field during the 1991 dry season. Each elementary plot (one seed lot) contained six rows, 5 m in length and separated by 0.75 m, which conforms to the spacing most commonly used by farmers in the study region. To obtain a sample representative of the diversity of each seed lot, seed for each plot was taken from 100 ears (two grains per ear) selected by the owner. Descriptors were measured using a sample of 20 plants and 15 ears per elementary plot, and refer to characteristics of the vegetative parts, tassel, and ear (see Table 5.2).

Factorial Discriminant Analysis (FDA) and Hierarchical Cluster Analysis (HCA) (STATITCF program) were used to analyze diversity among the seed lots within and across varieties. Factorial Discriminant Analysis distinguishes seed lots (or varieties) based on the variables for which the ratio of the sum of squared differences within a lot (or a variety) to the

Table 5.2 Vegetative and Ear Descriptors Measured

Vegetative descriptors	
HPL	Plant height
HEA	Ear height
DIA	Stalk diameter
LLE	Length of the leaf of the superior ear node
WLE	Width of the leaf of the superior ear node
NLE	Number of leaves above the superior ear, including the leaf of the superior ear node

Tassel descriptors	
LTA	Tassel length
PED	Peduncule length
LBR	Length of branched part of the tassel
BR	Total number of branches

Ear descriptors	
LEA	Ear length
WEA	Ear weight
DEA	Ear diameter
WCO	Cob weight
DCO	Cob diameter
ROW	Number of rows of grain
HGR	Grain height (3 grains mean)
WGR	Grain width (10 grains mean)
TGR	Grain thickness (10 grains mean)
W1G	1-grain weight (mean of 3 samples of 100 grains)

sum of squared differences among lots (or among varieties) is the greatest. Hierarchical Cluster Analysis ranks lots (or varieties) based on the mean of the weighted Euclidean distances among their center of gravity coordinates on the first five axes identified by the results of the FDA analysis. All variables were used in the FDA-HCA analyses except flowering date, grain color, and 1-grain weight obtained at the sample level (not at the plant level).

Phenotypic characteristics and varietal identification

With the exception of the Bl lot of the Blanco variety, the HCA analysis of seed lots for five of the more widely grown varieties (four locals and one exotic) demonstrates that seed lots bearing the same name cluster together based on their morphological characteristics (Figure 5.3). The results support the hypothesis that a farmer's concept of a variety corresponds closely to that of a phenotype. A farmer variety is a set of seed lots having the same name; these seed lots produce maize with similar plant, tassel, and ear characteristics.

The implication of these findings is that when farmers in Cuzalapa classify seed as that of a given variety, they use morphological and

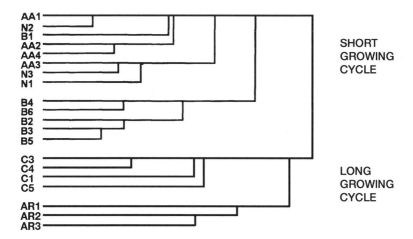

Figure 5.3 Hierarchical cluster analysis of seed lots of five varieties, by phenotypic characteristics (B = Blanco, AA = Amarillo Ancho, N = Negro, C = Chianquiahuitl, AR = Argentino).

phenological criteria rather than criteria such as geographic origin, adaptation to some limiting factor, or ritual function. A seed lot that resembles seed of a "local" landrace is classified as such by the farmer, even though its origin may be exotic or unknown. As a consequence, some seed lots of "local" landraces are in fact introduced from other regions.

Phenotypic variation between varieties

The phenotypic characteristics of the six local varieties and eight exotic cultivars (including the three most widely cultivated) were studied with the methods described above. The data reveal a large amount of phenotypic diversity with respect to several characters (Table 5.3). For example, the sum of degree days from sowing to tasseling varied from 1,130°C for the earliest maturing variety, Blanco, to 1,550°C for the latest maturing variety, Argentino. Mean height of the ear varied between varieties from 129 to 195 cm, the number of rows of grain varied from 8.7 to 12.7, the grain width from 0.85 to 1.13 cm, the cob diameter from 1.8 to 2.7 cm, and the ear weight from 104 to 181 g.

For the varieties studied, 78% of the variability in phenotypic characteristics was explained by the first two axes of the FDA (Figure 5.4). The first axis is essentially defined by row number (–ROW), grain width (+WGR), plant height (–HPL), and ear height (–HEA). The second axis is determined by ear development, including the weight and diameter of the cob (+WCO, +DCO) and weight and diameter of the ear (+WEA, +DEA). A test comparing farmers' methods for identifying varieties and these two axes indicated that both the statistical analysis and farmers classify maize varieties in a similar way (Louette 1994).

Table 5.3   Principal Characteristics of the 14 Varieties under Study (Descriptors in Table 5.2)

Varieties	MF day	HEA cm	HPL cm	NLE	LLE cm	RM	ROW	WGR cm	TGR cm	WCO g	DCO cm	PEA g	DEA cm	W1G g
Short cycle														
Blanco B	77.3	129	219	5.9	7.9	16.1	8.7	1.13	0.40	19.7	2.1	140	4.0	0.42
Perla P	82	144	235	6.1	8.1	16.9	8.7	1.08	0.39	18.7	2.2	128	3.9	0.38
Amarillo Ancho AA	82	146	231	6.1	7.9	19.3	9.8	1.00	0.39	19.8	2.2	126	3.9	0.33
Amarillo de Teq. AT	82	160	242	6.2	7.8	20.8	9.6	0.99	0.38	17.5	2.1	123	3.9	0.35
Negro N	83.2	156	232	6.3	7.9	19.8	10.0	0.97	0.37	18.1	2.2	123	3.9	0.31
Tabloncillo T	85	145	230	6.2	7.7	19.2	9.3	0.95	0.33	12.0	1.8	104	3.6	0.29
Long cycle														
HT47 HC	89.5	130	193	6.4	8.9	13.2	15.0	0.82	0.40	30.8	3.0	137	4.5	0.27
Negro (exot) NX	91.5	171	232	6.1	8.2	20.5	10.2	1.00	0.38	23.1	2.4	126	4.0	0.31
Híbrido H	92	179	254	6.3	8.1	20.4	11.9	0.91	0.37	22.0	2.3	141	4.2	0.30
Amarillo A	92	185	261	6.6	8.1	19.8	11.3	0.99	0.38	27.3	2.6	164	4.4	0.36
Enano E	92.5	161	231	6.8	8.5	23.2	13.4	0.89	0.40	29.7	2.7	160	4.5	0.31
Guino G	92.5	174	249	6.5	8.6	20.0	12.7	0.94	0.36	30.1	2.7	181	4.6	0.34
Chianquiahuitl C	93.2	188	260	6.2	7.8	21.5	11.7	0.85	0.34	17.6	2.1	126	3.9	0.27
Enano Gigante EG	93.5	185	261	6.6	8.4	20.5	12.4	0.93	0.36	26.2	2.6	158	4.4	0.32
Argentino AR	96	195	273	6.5	8.4	22.8	12.6	0.92	0.36	26.2	2.5	158	4.4	0.32

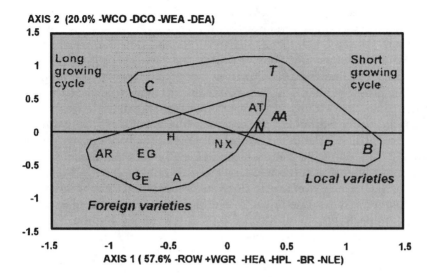

Figure 5.4 Phenotypic diversity in Cuzalapa maize (Factorial Discriminant Analysis).

The descriptors listed above facilitated the differentiation of varieties in two ways: by duration (length of growing cycle) and by origin or race. These characteristics could not be included as variables in the analysis because they were not collected at the level of the plant or ear, but at that of the seed lot. Nevertheless, they characterize well the different groups that appear in the FDA as they are closely related to some descriptors that define the first two axes of the FDA. Length of growing cycle is highly correlated with descriptors for the first axis ($|r| > 0.80$ between male flowering date and HEA, NLE, WGR, ROW). A long-duration variety is characteristically a taller plant that has more leaves and smaller grains arranged in more rows.

The origin of a variety (local or exotic) also relates to differences in phenotypic characteristics. The only exception to this general rule is the variety Amarillo [Tequesquitlán] (AT), which is phenotypically associated with the local varieties even though it was introduced from a community located some 20 km from Cuzalapa. The local varieties are characterized by narrower, lighter ears and less vegetative development than the exotic varieties (Table 5.3). Local varieties and Amarillo [Tequesquitlán] are related to the Tabloncillo race, which originated on the Pacific Coast of Mexico (Wellhausen et al. 1952). The exotic varieties included in the trial (except Amarillo [Tequesquitlán]) are linked to other races. Origin is therefore related to variation in race.

Origin and length of growing cycle are also interrelated. Most of the varieties with long growing cycles are exotic, with the exception of Chianquiahuitl. In Cuzalapa, therefore, local and exotic varieties appear to be complementary from a morphophenological point of view. Most local varieties have a short growing cycle, reduced vegetative development, few rows, and large kernels, whereas introduced varieties have a long growing cycle, taller plants, and small kernels.

There are three possible explanations for the fact that nearly all exotic varieties in Cuzalapa have long growing cycles, whereas local varieties have short ones. The first is that in Cuzalapa today, varieties with a short growing cycle are grown primarily in the dry season and long-cycle varieties are generally planted in the rainy season. Until the 1970s, flooded rice was cultivated during the rainy season and maize was sown almost exclusively during the dry season. The local landraces were then generally early maturing. The longer growing cycles of exotic varieties may reflect the fact that maize cultivation during the rainy season began on a large scale only recently.

Another explanation for the close relationship between the length of the growing cycle and exotic origin is that few landraces in the region around the Cuzalapa Valley mature early; outside Cuzalapa, the major cropping cycle for maize is the rainy season as most irrigated fields are sown with sugar cane. Few early maturing improved varieties have been developed for the lowland tropical zones where most maize is produced in developing countries (CIMMYT 1993).

Finally, the complementary characteristics of local and exotic varieties may be interpreted in yet a different way. When a lot of seed introduced from another community has the same phenotypic characteristics as seed of a local variety, farmers may consider it as seed of a local variety. The new seed would be identified by the name of the local variety and would no longer be distinguishable from it. For example, all introduced seed of maize with short, thick stalks is named Enano ("dwarf") after the first exotic variety that had such a stalk. Farmers do not use different names for these different varieties as the characteristic uses for classification refers to the height and diameter of stalk which are very similar among the different varieties. Farmers appear to use different names only for seed lots with particular characteristics of interest to them. Therefore, no introduced seed lot that is morphologically similar to a local variety would be distinguished, so no exotic variety with characteristics similar to those of local varieties would be recognized as a distinct cultivar.

Genetic definition of a landrace

Geneflow between seed lots

Monitoring geneflow

Maize is an open pollinated crop. If geneflow between local and exotic material is not controlled, the introduction of foreign varieties can have an important effect over the genetic structure of local ones. To evaluate the risk of geneflow between different varieties, we have studied over three seasons, on a 10-ha area, the sowing organization of seed lots in space (localization of the different seed lots) and in time (sowing date and flowering date). This area corresponded to seven fields separated from each other by less than 200 m. As this is the minimum distance for reproductive isolation in maize breeding (Hain-zelin 1988), geneflow can take place between all seed lots planted on this area.

In six farmers' fields, we evaluated the level of geneflow between seed lots sown on contiguous areas; using the xenia effect of the grain, and in particular the dominant character of the Negro color, an ovule with alleles that give white or yellow color to the grain, fecundated by a pollen with alleles that confer a purple or black color to the grain will give a purple or Negro grain at harvest time. The level of geneflow was then determined by the proportion of purple or Negro grains per furrow in the white or yellow varieties sown on contiguous areas to a Negro variety with a similar growing cycle. As the Negro variety is not homozygous for grain color, geneflow is probably greater than the one measured with the number of purple or Negro grains in the white or yellow variety.

Continual genetic exchange

The survey and the observation of the sowing pattern on an area of 10 ha during three cultivation seasons indicate that traditional management of seed lots does not aim to prevent the sowing in contiguous areas of different varieties (Figure 5.5). A farmer sows an average 2.5 varieties per cycle in the same field, independently of those sown on the contiguous fields. There is no physical isolation between local and exotic varieties and between locally reproduced seed lots and seed lots planted in other areas. For example, during the 1991 dry season, 15 seed lots, 3 of which were introduced from other regions, of six different varieties were sown in the area surveyed.

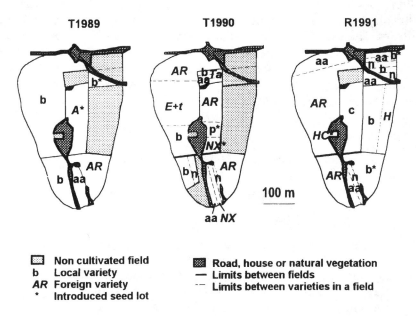

Figure 5.5   Location of maize varieties in seven fields, observed during three cropping seasons.

The planting date does not, however, lead to a sufficient difference of flowering date to permit reproductive isolation, in default of spacing isolation. In assessing the probability of geneflow between the different varieties sown, the work of Basseti and Westgate (1993) has shown that geneflow is more probable between two maize varieties when the difference between the male flowering date of a variety and the female flowering date of the other variety is less than 5 days. Over the three seasons observed in Cuzalapa, the differences of flowering dates between seed lots averaged less than 5 days in 38% of the cases. Different planting dates from farmer to farmer allowed for the synchrony of flowering between long cycle and short cycle varieties in 24% of the cases, although this situation has been more frequent for varieties with similar growing cycles (synchrony of flowering for 65% of the cases for long cycle varieties and for 47% of the cases for short cycle varieties).

Observation on six farmers' fields of the contamination of yellow- or white-grained varieties by a purple-grained variety planted in a contiguous area confirmed the presence of geneflow and provided insight into the level at which it occurs. It was observed, as indicated in the literature (Paterniani and Stort 1974), that the level of contamination of one variety by another diminished rapidly with distance from 20 to 10% in the first row to 1% after the first 2 or 3 m. The level stabilized over a great distance. The concentration of contamination in the first rows of contact between varieties may explain why some farmers think that contamination occurs at the root level.

The management of sowing practices, leading to the development of different varieties on contiguous areas, favors genetic exchange between all cultivar types, independent of the origin and growing cycle of the different varieties. The varieties sown are not genetically isolated. The reproduction of the varieties in the same conditions each cycle can lead to important modifications of their allelic frequencies. Thus, the genetic structure of local varieties is linked to the diversity of the varieties sown in the area and can be particularly influenced by exotic varieties.

Genetic drift

The study of the quantity of seed from which seed lots are reproduced provides evidence which confirms the genetic instability of local varieties and shows why geneflow between seed lots is so important in this system.

Determining the quantity of seed used per seed lot each cycle

Replanting each variety from small samples of seeds theoretically leads to a loss of alleles (Maruyama and Fuerst 1985; Ollitrault 1987). For an open pollinated plant, the theoretical work of Crossa (1988) and Crossa and Vencovsky (1994) has shown that a seed lot formed from less than 40 ears (1) does not permit the conservation of alleles whose frequency in the population in less than 3% (rare alleles), and (2) is conducive to the loss of heterocigosis superior to 1% when there are less than three alleles per locus. Thus, the use of reduced and variable quantity of seeds leads to the fluctuation of diversity with loss of alleles (Maruyama and Fuerst 1985; Ollitrault 1987).

To determine the effective population size of the seed lots planted in Cuzalapa, the volume of seed of each seed lot was obtained for the 39 farmers participating in the survey during six cultivation seasons. This was converted to the number of shelled ears for each variety based on the weight of one liter of grains and of 100 grains, and an average of 250 grains used for seed per ear.

Fluctuation of diversity

In Cuzalapa, as the field area is reduced (2 ha, on average) and various varieties are sown in the same field, the size of the seed lots planted per variety is reduced. More than 30% of seed lots sown during the six cultivation seasons covered by the survey were constituted from less than 40 ears (Figure 5.6). This phenomenon is important above all for varieties cultivated in small areas, such as the introduced varieties (37.7% of seed lots constituted from less than 40 ears) and the purple and yellow varieties (54.7%). For the main varieties, the phenomenon is less important, although more than 15% of the seed lots of the Blanco variety were constituted from less than 40 ears.

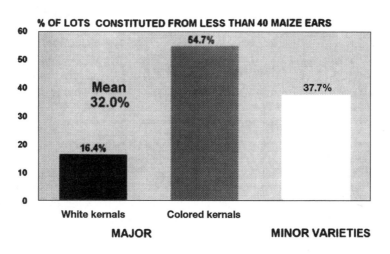

Figure 5.6 Quantity of maize ears shelled to the seed lots, by type of variety.

In conclusion, an important proportion of seed lots are submitted to a regular reduction of their effective population size, leading to the fluctuation of their diversity with loss of rare alleles. Similar findings have been reported by Ollitrault (1987) for rice, millet, and sorghum in Africa. If farmers managed seed lots in isolation from each other from a reproductive point of view, the diversity of some seed lots would probably decrease and consanguinity would probably increase, leading to a loss of production potential. In Cuzalapa, however, this is not the case. Consider, for example, that the genetic

Table 5.4 Isoenzymatic Polymorphism of Four Local Varieties

Variety	Sample (Number of Kernels)	Number of Alleles per Polymorphic Locus	Number of Rate Alleles (Frequency <5%)	Frequency of Polymorphic Locus	Genetic Diversity
Blanco	42	3.4 (0.7)	2	66.7	0.39
Amarillo Ancho	20	2.9 (0.6)	1	66.7	0.34
Negro	41	3.5 (0.3)	5	66.7	0.38
Chianquiahuitl	32	3.3 (0.3)	3	66.7	0.35

diversity of the Negro variety, reproduced from seed lots, of which 70% originate from less than 40 ears, is extremely similar to the diversity of varieties like Blanco, reproduced from significantly larger seed lots (Table 5.4). Geneflow is both responsible and necessary for the restoration of the genetic diversity of seed lots submitted to genetic drift.

Seed selection, conservation of phenotypic diversity, and control of geneflow

In this context, how can varieties maintain unique characteristics within a limited area? This polymorphism cannot be explained by limited geneflow compared to genetic drift. If this were the case, different seed lots of the same variety would tend to differentiate one from the other, which is not the case. Seed selection, in fact, seems to be part of the answer.

Farmers' seed selection

Determining seed selection criteria and the influence of selection over the genetic structure of varieties

The seed selection criteria used by farmers in the region were obtained from survey data and from a comparison, for five varieties, between the characteristics of samples of 60 to 140 ears selected by farmers for seed and samples of ears drawn at random from the harvest. As the genetic structure of an open pollinated plant such as maize can be modified by geneflow, an experiment was conducted to verify the extent to which seed selection permits the maintenance of characteristics in conditions of geneflow between seed lots. A random seed sample and a seed sample selected by a farmer were drawn from the farmer's harvest. Those samples were submitted to geneflow over two seasons, using a variety called a "contaminating variety" (Figure 5.7). Each seed sample was constituted from 100 ears and occupied an area of 20 × 20 m within the field planted with the contaminating variety. The initial population (R0) was compared to the last generation of seed selected (S2) and of seed drawn a random (R2), in a trial with four replications for their plant, tassel, and ear characteristics. They were also compared at the genetic level for 15

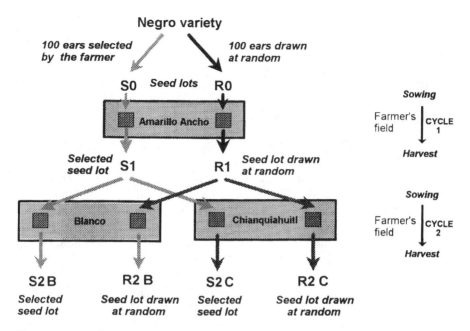

Figure 5.7   Method for determining the influence of seed selection over geneflow.

isoenzymatic loci: ACP-1, ACP-2, CPX-1, CPX-2, CPX-3, EST-8, GDH-2, GDH-3, GOT-1, GOT-2, IDH1-3, PGI-L, PGM-1, PGI-2, and SDH-1.

The trial conducted with Negro as the contaminated variety is considered in this chapter (Figure 5.7). In order to more accurately identify the effect of selection in relationship to the characteristics of the Negro variety, the experiment was subdivided during the second season. A pair of selected and nonselected seed lots were contaminated by the Blanco variety, which displays phenotypic and phenological characteristics similar to those of the Negro variety. Another pair of seed lots were submitted to contamination by the Chianquiahuitl variety, which has significantly different characteristics from the Negro variety (longer growing cycle, bigger vegetative development, more rows of grains, smaller grains, etc.). In this case the initial population (R0) was compared to the last one contaminated either by Blanco (S2B, R2B) or by Chianquiahuitl (S2C, R2C).

Seed selection criteria

Associated with the selection pressure of the environment, the seeds of varieties are mass selected by farmers. The selection does not exclude the ears produced in the borders of the field (with greater probability of contamination). It is based exclusively on the ear after harvest, without control of the pollen source or of the plant characteristics. The selected ears are, according to farmers, the well-developed ears with sane (without fungi or insect damage) and well-filled kernels, characteristics that correspond to the ideotype of the variety. From the selected ears, the kernels of the top of the ear, and sometimes those of the base as well, are not used as seed. These practices have been reported for other regions of Mexico, and for other countries (Bellon 1990; Johannessen 1982; SEP 1982).

The comparison between the characteristics of samples of ears selected for seed and samples of ears drawn at random for the Blanco, Amarillo Ancho, Negro, Chianquiahuitl, and Argentino varieties shows that important differences exist between those two types of ears (Figure 5.8). Selection is oriented to the well-developed ears: axis 2 of the Factorial Discriminant Analysis (FDA) is determined by the weight (WEA) and diameter (DEA) of the ear, the weight of the cob (WCO), and the height of the kernel (HGR). The differences are generally very significant. For example, the mean weight of the selected ears is 30% higher than the mean weight of the unselected ears. As seed selection is oriented to the more developed ears, selection favors the more productive and/or adapted genotypes. This allows for the maintenance of productive varieties. For varieties of a different growing cycle, the selection is divergent on the first axis of the FDA. This axis is related to the characteristics of the ear that distinguish varieties of short and long growing cycle length: number of grain rows (ROW), width (WGR), and thickness (TGR) of the grain. The differences are not always statistically different, although the tendency is evident in the FDA (Figure 5.8) and on the value of the descriptors. In Cuzalapa, the seed selection strengthens the characteristics that distinguish the varieties according to their growing

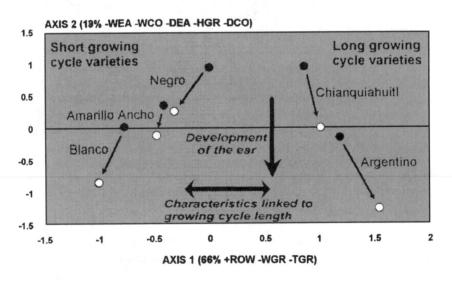

Figure 5.8 Seed selection criteria for five of the main varieties (Factorial Discriminant Analysis).

season. From an agroecological perspective, this has important implications for maize production in a region with two growing seasons.

Control of geneflow for ear characteristics

Comparing traditional seed selection with the random selection of seed in conditions of contamination by another variety shows that the traditional selection of seed efficiently conserves the characteristics of the ear. Let us consider the trial in which the Negro variety (Figure 5.7) was contaminated during the second season by the Blanco variety (similar to the Negro variety from a phenological and phenotypical point of view) and the Chianquiahuitl variety (different from a phenotypical point of view, and having a longer growing cycle; Table 5.3). We compare the characteristics of the initial population (R0) and the population resulting from the two seasons of contamination: populations drawn at random and contaminated by Blanco (R2B) or Chianquiahuitl (R2C), and populations selected for seed and contaminated by Blanco (S2B) or by Chianquiahuitl (S2C).

As shown by the relative position of R0 and R2B in the FDA (Figure 5.9), the contamination by the Blanco variety, similar to the Negro variety, has had little effect on the Negro variety. In contrast, the relative position of R0 and R2C indicate that the contamination by Chianquiahuitl, different from Negro, has led to some modifications in the Negro variety. The changes that occurred are related to the characteristics of the Chianquiahuitl variety. R2C presents a vegetative development superior to that of R0. The first axis is defined by the height of the plant (HPL) and of the ear (HEA), the number of leaves (NLE) and the diameter of the stalk (DIA) and smaller kernels arranged on more rows (the first axis is defined by the weight of the kernels, (WG), the width of the grain (WGR) and the number of grain rows (ROW). The values are statistically different only for DIA.

The selection of the seed has had the same effect on the contaminated populations for both contaminating varieties: a higher vegetative development, as indicated by the relative position of S2B and S2C on the first axes of the FDA. What is more interesting to note is that the selection of the population contaminated by Chianquiahuitl has led to the reduction of the effect of contamination over the characteristics of the ears as shown by the relative positions of S2C and R2C on the second axes. The values are statistically different between R2C and S2C for the width of the grain (WGR) and for the number of rows of grain (ROW), which are precisely the characteristics used by farmers to select their seeds.

The effect of selection has also been documented over other characteristics important to farmers when selecting seed: the color of the grain. Over the two growing seasons, the Negro variety was submitted to contamination by white or yellow varieties. The ratio of white or yellow grains in the Negro variety has increased from 7.5 to 16.5% when seed was drawn at random while it stood stable when seed was selected.

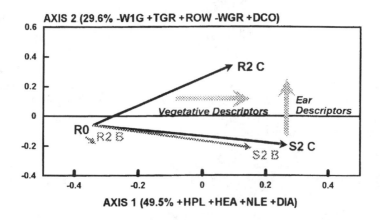

R0	INITIAL POPULATION of the Negro variety	**2d GENERATION CONTAMINATED BY:**			
		B	Selected S2B At random R2B	**C**	Selected S2C At random R2C

Figure 5.9   Contamination of the Negro variety by Blanco and Chianquiahuitl and contamination control by seed selection.

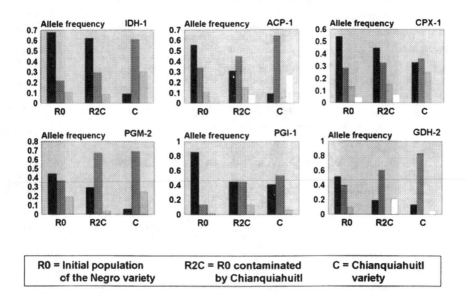

R0 = Initial population of the Negro variety	R2C = R0 contaminated by Chianquiahuitl	C = Chianquiahuitl variety

Figure 5.10   Genetic effect of the Negro variety contamination by the Chianquiahuitl variety.

At the genetic level, we were able to observe the effect of contamination on six of the ten polymorphic loci examined (Figure 5.10). The population without selection (R2C) has frequencies intermediate between those of the initial population (R0) and those of the contaminating variety Chianquiahuitl (C). Nevertheless, no effect of selection could be observed at the genetic level.

Discussion and conclusions

Cuzalapa as an open system from the germplasm point of view

The study of seed exchange and the morphological structure of diversity suggests that on-farm conservation projects must not isolate a rural community or a variety. The assumption that traditional systems are closed and isolated with respect to the flow of genetic material is clearly contradicted by the results of this study. A small group of local landraces is continuously cultivated, while varieties with diverse origins that are morphologically diverse among themselves and distinct from the local landraces succeed each other over time. These exotic varieties are introduced for testing by farmers, but they may also be integrated into the group of local landraces. This case study shows that over 3 years alone, in a traditional farming system located in what some regard as the geographic center of origin for maize, introduced materials represent a substantial proportion of the maize seed planted and local varieties are not generally the product of exclusively local seed selection and management.

Similar results have been reported by researchers investigating the use of rice (Dennis 1987), maize (Bellon and Brush 1994; Ortega 1973), bean (Sperling and Loevinsohn 1993), and potato varieties (Brush et al. 1981). Dennis (1987) and Bellon (1995) characterize similar situations as an "excess of diversity" with respect to what is necessary to keep the agricultural system functioning.

Rather than displacing local cultivars, exotic varieties occupy a small proportion of the area planted to maize, and local landraces continue to dominate maize area in Cuzalapa. Introduced varieties more often have uses and modes of management that are complementary, rather than substitutable for, those of the dominant cultivars (Bérard et al. 1991). Our findings in Cuzalapa are similar to what Bellon and Brush (1994) described in Chiapas, although the proportion of local and exotic maize varieties is reversed in the two cases.

The appropriate geographic scale over which we can define a variety as "local" is problematic because the mechanisms that explain the phenotypic diversity of maize in Cuzalapa suggest a constant influx of genetic material. Exotic varieties, as well as introduced seed lots that are then integrated into local varieties, are probably a source of phenotypic and genetic diversity. It is questionable whether any particular geographic scale would necessarily include all of the factors affecting "local" varieties.

In Cuzalapa the morphophenological characteristics of the local and exotic varieties seem complementary, and the two groups rarely compete with respect to growing cycle, vegetative characteristics, or ear attributes. Introductions do not necessarily lead to a large shift away from local cultivars. This finding suggests that a variety is more easily adopted by farmers if it satisfies a need that is not currently met by local varieties or if it occupies a place in the morphological continuum that has not yet been exploited (Boster 1985). In Cuzalapa, survey farmers clearly sought new or different genetic materials from among exotic varieties.

At the level of introduction observed in Cuzalapa, exotic varieties are more a source of phenotypic diversity than a factor inducing genetic erosion. As indicated by Brush (1992), genetic erosion seems to be a phenomenon that is too complex to be captured in the equality "introduction of varieties = loss of genetic diversity." Genetic erosion is a complex function of the area occupied by introductions vs. area planted to local cultivars, the diversity within and between the introduced and local cultivars, and the extent to which local varieties have been abandoned or displaced. As long as the function of the introduced material is complementary to that of the local germplasm, diversity probably increases. When introduced and local materials compete, exotic varieties can displace local material, but this displacement leads to a loss of diversity only if the introduced material is less diverse, replaces several local landraces, or displaces genetic uniqueness of local material. Identifying the factors that affect the extent of genetic erosion, and determining their critical values, is likely to be difficult, and especially so in a system as dynamic as that of Cuzalapa.

The regular acquisition of genetic material by farmers is evidence of their interest in, rather than resistance to, the introduction of new cultivars. In Cuzalapa, farmers are generally experimenters who do not hesitate to test new cultivars planted by farmers in other regions against their respective local varieties. Brush et al. (1981) have indicated that in the Mantaro Valley in Peru farmers may travel more than 50 km in search of new potato varieties. Farmers in Cuzalapa will adopt a maize variety, however, only if it demonstrates its advantages consistently over a large number of cropping seasons. One unsuccessful trial can lead a farmer to abandon a variety, regardless of the reason for the failure. In Dennis' (1987) study in Thailand, farmers in the eight survey villages, on average, cultivated ten varieties in the first year, adopted four introduced varieties in 5 years, and abandoned four cultivars during the same period. Over the past 40 years in Cuzalapa, of all of the varieties introduced by the survey farmers of Cuzalapa, only Chianquiahuitl has been fully adopted.

A local variety as an open genetic structure

Another major research finding concerns the definition of a local variety itself. The magnitude of seed exchange among farmers and the fluctuation of the diversity of seed lots, caused by the amount of seed used and by the regular geneflow between seed lots, raise questions about farmers' concepts of a variety and the distinction between "local" and "exotic" varieties.

First of all, in Cuzalapa, it is not only the set of cultivars but also the set of seed lots that constitute the cultivars which vary in time. A certain number of seed lots disappear in each crop cycle because they are not replanted by the farmer who selected them; on the other hand, one introduced seed lot may evolve into a number of seed lots once farmers begin to exchange seed. The composition of the group of seed lots that constitute a variety is mutable over time. The geographical point of reference for the term "local variety" is larger than the community itself. Introduced seed lots that phenotypically resemble seed of local landraces are adopted as part of these landraces. Thus, the genetic diversity of a variety can be traced beyond the community itself. No geographic scale can exactly define a variety.

Finally, seed lots are submitted to fluctuations in their levels of diversity due to the changing amount of seed from which they are reproduced and continuous geneflow from other seed lots. A farmer variety is, therefore, mutable in terms of the number, origin, and genetic composition of the seed lots of which it is composed. Contrary to the modern concept of variety, traditional cultivars are not genetically stable populations that can be well defined for conservation purposes. Rather, local varieties constitute systems that are genetically open.

Seed selection for maintaining productivity and morphological characteristics

The traditional selection described in Cuzalapa has several utilitarian functions: to maintain the agronomic characteristics of the varieties by selecting the best ears, to maintain distinct morphological characteristics by selection based on those criteria, and to maintain diversity when the pollen source is not controlled (Sandmeier et al. 1986). Although the effect of selection over the conservation of phenotypic characteristics is not as strong as the effect of selection over agronomic characteristics, it seems systematic and has been demonstrated both by the experimental results and by statements of farmers. Traditional seed selection seems, therefore, to be an efficient means of conserving the integrity of the ear characteristics even when geneflow is a significant factor. As indicated by Boster (1985), varieties must be distinct in order to be selected for more utilitarian characteristics. If a variety is not easily distinguishable at the moment of seed selection, it can be lost in favor of varieties sown on more extensive areas. In Cuzalapa, seed selection facilitates the conservation of differences between varieties that have distinct functions within the area, particularly varieties of different growing cycles, length, or grain color. Seed selection does not, however, control geneflow that affects the characteristics of the ears at the genetic level or for vegetative characteristics. Therefore, traditional seed selection conserves phenotypic characteristics of the ear, but not the genetic integrity of the different seed lots.

Metapopulation structure

In Cuzalapa, what do farmers conserve? What is a landrace? What should be considered as the unit of conservation? The structure and processes described for maize cultivation in Cuzalapa can be compared to a meta-population structure, defined as a group of subpopulations interconnected by geneflow and submitted to local colonization and replacement by new populations (Olivieri and Gouyon 1992; Slatkin and Wade 1978) (Figure 5.11). In the case of Cuzalapa, the metapopulation is integrated into the various seed lots of the different varieties sown, through geneflow. The phenomenon of replacement corresponds to the disappearance of seed lots when they are not replanted, and the phenomenon of colonization corresponds to the creation of new seed lots through the interchange of seed between farmers. The maize metapopulation in Cuzalapa is interconnected with other metapopulations as seed lots are introduced from other regions and seed from Cuzalapa is sown in other areas.

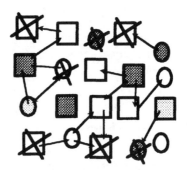

□ O	Seed lot of local variety (more than 40 ears / less than 40 ears)	
▓ ◉	Seed lot of foreign variety (more than 40 ears / less than 40 ears)	
⟶	Geneflow	
✗	Seed lot not reproduced during the next cycle	
▨ O	Seed lot introduced from another region	

Figure 5.11   Metapopulation in Cuzalapa.

Based on the various models which have elaborated on metapopulation structures (David 1992; Dickinson and Antonovics 1973; Hedrick 1986; Michalakis and Olivieri 1993; Nagylaki 1976; Roof 1994; Slatkin 1981; Slatkin 1989; Varvio et al. 1986; Zhivotovsky and Feldman 1992), we can interpret the genetic functioning of this metapopulation. First of all, this structure warrants the conservation of the global allelic diversity (Varvio et al. 1986). For example, in

Cuzalapa, as a variety is represented by several seed lots, an allele which is lost when selecting a seed lot from an insufficient number of ears (genetic drift with loss of alleles) can be retained in one or more other seed lots selected by other farmers and then conserved, at least temporarily, at the watershed scale. Geneflow between seed lots can then restore the genetic diversity of the seed lots submitted to genetic drift. Furthermore, Michalakis and Olivieri (1993) and Slatkin (1989) have shown that the effect of geneflow is reinforced by the phenomenon of colonization and replacement.

Analysis of metapopulation structures has also shown that it favors a dynamic evolution of diversity. Genetic drift and the introduction of new varieties favor the appearance of new genetic structures that are impossible to obtain in a unique population in panmixis and geneflow allows them to spread (Slatkin 1981). The management of varieties in such a system permits introduced varieties to serve as new material for the local varieties. Slatkin (1981) has shown that in a metapopulation structure, recessive or subdominant alleles respond better to selection than in a unique population in panmixis, the contrary being true for dominant alleles. That is, even if dominant alleles are selected, they have less potential to dominate. Likewise, the proportion of the recessive or subdominant alleles is easier to increase than in a population in panmixis. By permitting the permanence of all types of alleles, this structure warrants some level of diversity that can be considered "useless" at present, but which may prove important for the continuous adaptation of varieties.

Finally, in a metapopulation structure, polymorphism is favored by variable selection pressure over different subpopulations and reduced geneflow between them (David 1992; Dickinson and Antonovics 1973; Hedrick 1986; Nagylaki 1976; Roof 1994; Zhivotovsky and Feldman 1992). Strong geneflow — relative to selection pressures — would lead to uniformity over the set of subpopulations and reduction of global diversity, while the absence of geneflow would lead to inbreeding and to the death of some subpopulations. The phenotypic integrity present in Cuzalapa is maintained by different farmer selection criteria for different varieties, and by reduced geneflow.

While seed selection conserves the phenotypic integrity of the varieties, the processes occurring within the metapopulation structure formed by the group of varieties sown suggest that landraces are genetically variable over time. The traditional management of maize in Cuzalapa contributes more to the conservation of a general level of diversity than to the conservation of genetically stable and distinct maize populations. A landrace is far from a stable, distinct, and uniform unit. Its diversity is linked to the diversity of the material sown in the area, and then related to the diversity of the introduced varieties.

Implications for in situ conservation

The characterization of the maize farming system in the Cuzalapa watershed as open with respect to genetic material contrasts with the original model for conserving crop genetic resources *in situ*, in which farmers would be motivated not to change their cultural practices or introduce exotic genetic material (Altieri and Merrick 1987; Iltis 1974). A farming system is affected by exchange with other communities, and a variety is the product of genetic exchange with materials that may or may not be replanted locally.

Conservationists may argue that if a community under study reveals these characteristics, it is not traditional, because traditional systems are autarchic. In fact, the characterization of a society or community is normative and relative: a community is traditional only with respect to what is perceived as modern and with respect to other contemporary human groups. In any case, the system of seed exchange that has been described by farmers and observed in Cuzalapa appears "traditional" in the sense that it is customary and long-lived. With time and improved communication with other regions, the level of seed exchange might have changed, but not the interest in looking for new genetic materials. It is likely that the major findings reported can be generalized to other rural areas of Mexico, because the factors that explain seed exchange system in Cuzalapa appear neither new nor specific to this region. To be convinced of this point, it is enough to observe the extent to which the world is the fruit of an ancient and continuous evolution that includes the diffusion of plants from their centers of domestication, the adoption and abandonment of cultivars or of cultivated plants, the differentiation of races and varieties within species, and the adaptation of cultivars to various agrosystems and techniques of cultivation (Harlan 1992; Haudricourt and Hedin 1987).

This study has shown that the set of seed lots that constitute a variety and its diversity is mutable in time. The seed exchange between farmers and the geneflow between seed lots implies that varieties evolve within the entire set of genetic material planted in the region. A seed lot does not evolve as a specific farmer line. As Berg stated (1992), it has become clear that the proper conservation unit is not a variety, and never one single seed lot per variety, but the group of cultivated varieties in their subdivision and mixture.

The diversity found in this region is the fruit of collective management of local and exotic varieties. Although individual farmers cultivate several varieties, they cannot maintain the processes that support regional diversity in isolation from other farmers. Therefore, we must focus on the mechanisms that influence the metapopulation formed by all exotic and local varieties. What is important to preserve is not the genetic material in and of itself, but the processes that create and preserve genetic diversity.

Finally, what is the significance of on-farm conservation of local varieties and what are the optimum tools for implementing on-farm conservation strategies? Is the term "conservation" appropriate? There is no single answer. Rather, the answer depends on the objective of on-farm conservation, as well as the definition of the diversity to be conserved. On this topic, the positions are not clear. In most cases, the objectives of on-farm and *ex situ* conservation are considered the same. There is a lack of debate about the role of on-farm conservation in relation to the efforts of genetic resources conservation. In

Cuzalapa, for example, it seems clear that we cannot expect complete preservation of the genetic diversity actually present in the watershed. In this case, the conservation of the material in a gene bank would be a more appropriate option, provided that appropriate methods are used to collect samples. Equally, *ex situ* conservation would be the best alternative if the objective is to conserve specific alleles.

If conservation of the phenotypic characteristics of the local varieties is considered to be the objective of an on-farm conservation project in the zone, it would be sufficient to sow the varieties on areas of adequate size to reduce genetic drift and to ask farmers to select the seed. This material can alternately be sown in farmers' fields and conserved in an official or community gene bank. If the objective is to conserve the characteristics related to environmental adaptation of this material, diverse varieties could succeed one to the other if cultivated long enough in the zone to be locally adapted, acquiring these characteristics by geneflow or environmental selection. New cultivars could also be produced from the local ones (Oldfield and Alcorn 1987).

One could also ask if it is both realistic and necessary — for world agriculture and the development of the Cuzalapa community or its agro-system — that the Blanco, Amarillo Ancho, Perla, Negro, and Chianquiahuitl varieties are cultivated during the next century? Perhaps what is more important than the preservation of these varieties is the maintenance of a high level of phenotypic and genetic diversity: assuring that 20 different varieties continue to be cultivated in Cuzalapa, though not necessarily the same ones, and ensuring a high level of diversity for the introduced material, as massive introduction of varieties with low genetic diversity can lead to a reduction of the overall diversity. In this way, we turn the discussion from on-farm conservation of varieties to on-farm conservation of diversity.

References

Altieri, M.A. and L.C. Merrick. 1987. *In situ* conservation of crop genetic resources through maintenance of traditional farming systems, *Economic Botany* 41:86–96.

Basseti, P. and M.E. Westgate. 1993. Senescence and receptivity of maize silks, *Crop Science* 33:275–278.

Bellon, M.R. 1990. *The Ethnoecology of Maize under Technological Change*. Ph.D. Dissertation, University of California, Davis. University Microfilms, Ann Arbor, Michigan.

Bellon, M.R. 1995. The dynamics of crop infraspecific diversity: a conceptual framework at the farmer level, *Economic Botany* 50:26–39.

Bellon, M.R. and S.B. Brush. 1994. Keepers of maize in Chiapas, Mexico, *Economic Botany* 48:196–209.

Benz, B.F. 1988. *In situ* conservation of the genus *Zea* in the Sierra de Manantlán Biosphere Reserve. In *Recent advances in the conservation and utilization of genetic resources: Proceedings of the Global Maize Germplasm Workshop*. Mexico D.F: CIMMYT.

Benz, B.F. n.d. On the origin, evolution and dispersal of maize. In *The Beginnings of Agriculture and Development of Complex Societies: The Prehistory of the Pacific Basin*, M. Blake (ed.). Pullman, WA: Washington State University Press.

Benz, B.F. and H.H. Iltis. 1992. Evolution of female sexuality in maize ear (*Zea mays* L. subsp, *mays* Gramineae), *Economic Botany* 46:212–222.

Benz, B.F., L.R. Sánchez V., and J.F. Santana M. 1990. Ecology and ethnobotany of *Zea diploperennis*: preliminary investigations, *Maydica* 35:85–98.

Bérard, L., A. Fragata, A. de Carvalho, P. Marchenay, and J. Vieira da Silva. 1991. Cultivars locaux, ethnobiologie et developpement. In *La conservation des especes sauvages progénitrices des plantes culfivées. Collection Rencontres Environnement*, N. 8. Actes du colloque organisé par le Conseil de l'Europe en collaboration avec l'Office des Réserves Naturelles d'Israbl. Conseil de l'Europe, Strasbourg, France.

Berg, T. 1992. Indigenous knowedge and plant breeding in Tigray, Ethiopia, *Forum for Development Studies* 1:13–22.

Bommer, D.F.R. 1991. The historical development of international collaboration in plant genetic resources. In *Searching for new concepts for collaborative genetic resources management: Papers of the EUCARPIA/IBPGR Symposium*, edited by Th.J.L. van Hintun, L. Frese, and P.M. Perret. Wageningen, The Netherlands, 3–6 Dec. 1990, International Crop Networks Series N. 4. Rome: IPGRI.

Boster, J.S. 1985. Selection of perceptual distinctiveness: Evidence from Aguaruna cultivars of *Manihot esculenta*, *Economic Botany* 39:310–325.

Brush, S.B. 1991. A farmer-based approach to conserving crop germplasm, *Economic Botany* 45:153–165.

Brush, S.B. 1992. Farmer's rights and genetic conservation in traditional farming systems, *World Development* 20:1617–1630.

Brush, S.B., H.J. Carney, and Z. Huaman. 1981. Dynamics of Andean potato agriculture, *Economic Botany* 35:70–88.

CIMMYT (Centro lnternacional de Mejoramiento de Maiz y Trigo). 1993. *Genetic resources preservation, regeneration, maintenance and utilization. Briefing book*. Mexico, D.F: CIMMYT.

Cohen, J.I., J.B. Alcorn, and C.S. Potter. 1991. Utilization and conservation of genetic resources: International projects for sustainable agriculture, *Economic Botany* 45:190–199.

Cooper, D., R. Velvee, and H. Hobbelink (eds). 1992. *Growing Diversity: Genetic Resources and Local Food Security*. London: IT Publication, GRAIN.

Crossa, J. 1988. Theory and practice in determining sample size for conservation of maize germplasm. In *Recent advances in the conservation and utilization of genetic resource: Proceedings of the global maize germplasm workshop*. Mexico D.F.: CIMMYT.

Crossa, J. and R. Vencovsky. 1994. Implication of the variance effective population size on the genetic conservation of monoecious species, *Theoretical and Applied Genetics* 89:936–942.

David, J. 1992. *Approche méthodologique d'une gestion dynamique des ressources génétiques chez le blé tendre (Triticum aestivum L.)*. Thèse de doctorat en sciences, INA-PG.

Dennis, J.V., Jr. 1987. *Farmer Management of Rice Variety Diversity in Northern Thailand*. Ph.D. dissertation, Cornell University. Ann Arbor, MI: University Microfilms.

Dickinson, H. and J. Antonovics. 1973. Theoretical considerations of sympatric divergence, *The American Naturalist* 107:256–274.

FAO (Food and Agriculture Organization of the United Nations). 1989. *Ressources phytogénétiques: Leur conservation in situ au service des besoins humains*. Rome: FAO.

Hainzelin, E. 1988. *Manuel du producteur de semences de maïs en milieu tropical*. 30 questions-réponses élémentaires. France: IRAT/CIRAD.

Harlan, J.R. 1992. *Crops and Man*. Madison, WI: American Society of Agronomy and Crop Science Society of America.

Haudricourt, A.G. and L. Hedin. 1987. *L'homme et les plantes cultivées*. Paris: A.-M. Métaillié.

Hedrick, P.W. 1986. Genetic polymorphism in heterogeneous environment: a decade later, *Annual Review of Ecological Systems* 17:535–566.

Hernández X., E. 1985. Maize and man in the greater southwest, *Economic Botany* 39:416–430.

Iltis, H.H. 1974. Freezing the genetic landscape: The preservation of diversity in cultivated plants as an urgent social responsibility of plant geneticist and plant taxonomist, *Maize Genetics Cooperation News Letter* 48:199–200.

Jardel P.E. (Coord.). 1992. *Estrátegia para la conservación de la Reserva de la Biosphera Sierra de Manantlán*. Laboratorio Natural Las Joyas, Editorial Universidad de Guadalajara, Guadalajara, Jalisco, Mexico.

Johannessen, C.L. 1982. Domestication process continues in Guatemala, *Economic Botany* 36:84–99.

Keystone Center. 1991. *Final consensus report: Global initiative for the security and sustainable use of plant genetic resources*. Third Plenary Session, 31 May–4 June 1991. Keystone International Dialogue Series on Plant Genetic Resources, Oslo, Norway.

Laitner, K. and B.F. Benz. 1994. *Las condiciones culturales y ambientales en la reserva de la biosfera Sierra de Manantlán en tiempo de la conquista: una perspective de los documentos etnohistóricos secundarios*. Estudios del hombre, Universidad de Guadalajara, Guadalajara, Mexico. 1:15–45.

Louette, D. 1994. *Gestion Traditionnelle de variétés de maiz dans la réserve de la Biosphere Sierra de Manantlán (RBSM, états de Jalisco et Colima, Mexique) et conservation in situ des ressources génétiques de plantes cultivées*. Thèse de doctorat, Ecole Nationale Supérieure Agronomique de Montpellier, Montpellier, France.

Martínez, R. and J.J. Sandoval L. 1993. Levantamiento taxonómico de suelos de la subcuenca de Cuzalapa, Sierra de Manantlán, JAL, *Terra* 11:3–11.

Martínez, R., J.J. Sandoval L., and R.D. Guevara G. 1991. Climas de la reserva de la biosfera Sierra de Manantlán y su Area de influencia, *Agrociencia* 2:107–119.

Maruyama, T. and P.A. Fuerst. 1985. Population bottlenecks and nonequilibrium models in population genetics. III. Genic homozygosity in populations which experience periodic bottleneck, *Genetics* 111:691–703.

Merrick, L.C. 1990. Crop genetic diversity and its conservation in traditional agroecosystems. In *Agroecology and Small Farm Development*. M.A. Altieri and S.B. Hecht (eds.). Boca Raton, FL: CRC Press.

Michalakis, Y. and I. Olivieri. 1993. The influence of local extinctions on the probability of fixation of chromosomal rearrangements, *Journal of Evolutionary Biology* 6:153–170.

Montecinos, C. and M.A. Altieri. 1991. *Status and Trends in Grass-Roots Crop Genetic Conservation Efforts in Latin America*. CLADES and University of California at Berkeley.

Nagylaki, T. 1976. Dispersion-selection balance in localised plant populations, *Heredity* 37:59–67.

Oldfield, M.L. and J.B. Alcorn. 1987. Conservation in traditional agroecosystems, *Bioscience* 37:199–208.

Olivieri, I. and P.-H. Gouyon. 1992. Evolution des métapopulations et biodiversité. In *Actes du colloque nternabonal en hommage a Jean Pernes. Complexes d'especes, flux de génes et ressources génétiques des plantes*, 8–12 janvier 1992, Bureau des Ressources Génétiques, Paris, France.

Ollitrault, P. 1987. *Evaluation génétique des sorghos cultivés (Sorghum bicolor M. Moench) pour l'analyse conjointe des diversités enzymatique et morphophysiologiue. Relation avec les sorghos sauvages.* Thèse de doctorat Université Paris XI, Centre Orsay.

Ortega P., R.A. 1973. *Variación en maíz y cambios socioeconómicos en Chiapas, México 1946–1971.* Tesis de M.C. especialidad en Botánica, Colegio de Postgraduados, Chapingo, México.

Paterniani, E. and A.C. Stort. 1974. Effective maize pollen dispersal in the field, *Euphytica* 23:129–134.

Roff, D.A. 1994. Evolution of dimorphic traits: effect of directional selection on heritability, *Heredity* 72:36–41.

Rosales, A.J.J. and M.S.H. Graf. 1995. *Diagnóstico sociodemográfico de la Reserva de la Biosfera Sierra de Manantlán y su región de influencia.* IMECIO, Universida de Guadalajara, El Grullo, informe técnico.

Sandmeier, M., S. Pilate-Andre, and J. Pernes. 1986. Relations génétiques entre les populations de mils sauvages et cultivés: résultats d'une enquête au Mali, *Journal d'Agriculture Traditionnelle et de Botanique Appliquée* 33:69–89.

SEP - Secretaría de Educación Pública. 1982. *Nuestro maíz, treinta monografías populares.* Consejo nacional de Fomento Educativo, Secretaría de Educación pública, Tomo 1 y 2, Mexico.

Slatkin, M. 1981. Fixation probabilities and fixation times in a subdivided population, *Evolution* 35:477–488.

Slatkin, M. 1989. Population structure and evolutionary progress. *Genome* 31:196–202.

Slatkin M. and M.J. Wade. 1978. Group selection on a quantitative character, *Science* 75:3531–3534.

Sperling, L. and M.E. Loevinsohn. 1993. The dynamics of adoption: distribution and mortality of bean varieties among small farmers in Rwanda, *Agricultural Systems* 41:41–453.

Varvio, S.-L., R. Chakraborty, and M. Nei. 1986. Genetic variation in subdivided population and conservation genetics, *Heredity* 57:189–198.

Wellhausen, E.J., L.M. Roberts, and E. Hernández X. with P.C. Mangelsdorf. 1952. *Races of Maize in Mexico: Their Origin, Characteristics and Distribution.* Cambridge, MA: Harvard University, Bussey Institution.

Zhivotovsky, L.A. and M.W. Feldman. 1992. On models of quantitative genetic variability: a stabilizing selection-balance model, *Genetics* 130:947–955.

chapter six

Keeping diversity alive: an Ethiopian perspective

Melaku Worede, Tesfaye Tesemma, and Regassa Feyissa

Abstract

In Ethiopia, a region representing a major world gene center, the various traditional agroecosystems constitute major *in situ* repositories of crop genetic diversity. Maintenance of species and genetic diversity in the field is one of the effective strategies whereby resource-poor farmers practice low-input agriculture in marginal environments to create stable systems. The existence of such native germplasm is also crucial to sustained provision of useful genetic material to breeding programs worldwide. This chapter describes the role of biodiversity in Ethiopian agriculture and also highlights the various factors that account for the maintenance of diversity on peasant farms. The major threats of loss of genetic diversity are discussed, particularly in the context of agricultural modernization and environmental degradation now in progress in Ethiopia.

Genetic resource activities represent a major national effort that Ethiopia has undertaken systematically over the past two decades. The existing options pose a serious challenge to the country, requiring major inputs in terms of technical know-how and material. There is also a unique opportunity to conserve landraces in a dynamic state, on peasant farms. In this context, we describe past and present activities of the Seeds of Survival Program/Ethiopia (SoS/E), a participatory, dynamic, farmer-based approach to crop genetic resource (landrace) conservation, enhancement, and utilization (on-farm multiplication and distribution).

Introduction

Ethiopia is a major world center of genetic diversity for many important domesticated crop plant species such as sorghum, barley, teff, chickpeas, and coffee, largely represented in the country by landraces and wild types that are uniquely adapted, genetically diverse forms of these various crops. The genetic diversity found in Ethiopian landraces has been used worldwide in developing new crop varieties and addressing acute yield constraints. Much of this crop diversity is found in small fields of peasants who, aided by nature, have played a central role in the creation, maintenance, and use of these invaluable resources. Peasant farmers in Ethiopia translate their deep understanding and use of different plants and animals, or the general biology of their surroundings, to farming systems that are best adapted to their own circumstances.

The existence of genetic diversity has special significance for the maintenance and enhancement of productivity in agricultural crops in a country like Ethiopia, which is characterized by highly varied agro-climates and diverse growing conditions. Such diversity provides security for the farmer against diseases, pests, drought, and other stresses. Genetic diversity also allows farmers to exploit the full range of the country's highly varied micro-environments differing in characteristics such as soil, water, temperature, altitude, slope, and fertility. Diversity among species is especially significant to Ethiopia as it represents an important resource to subsistence farming communities throughout the country. A wide variety of plant and animal species provide material for food, feed, fiber, and medicinal uses. Such diversity is also crucial to sustain current production systems, improve human diets, and support biological systems essential for the livelihood of local communities. Maintenance of species and genetic diversity in farmers' fields is, therefore, crucial to sustainable agriculture, especially for resource-poor farmers practicing agriculture under low-input conditions in marginal lands.

In this chapter, we describe the farmer-based approach to conservation and use of crop genetic materials that the Seeds of Survival Program/Ethiopia (SoS/E) has undertaken in partnership with the former Plant Genetic Resources Center/Ethiopia (PGRC/E), now Ethiopian Biodiversity Institute (EBI), since 1988 as a Unitarian Service Committee of Canada (USC/C) supported program. This chapter discusses the cooperative role that farmers, scientists, and local extension agents play to make the program relevant to food and livelihood security of small-scale farmers, as well as growing urban populations. The importance of crop genetic diversity in sustaining productivity and the livelihoods of resource-poor farmers working under adverse growing conditions and the risk of genetic erosion are also discussed.

Farmers: key players in sustaining diversity

In Ethiopia, traditional farming represents centuries of accumulated experience and skills of peasants who often sustained yields under adverse farming conditions using locally available resources. The foundation for Ethiopian farming is comprised of the traditional crops and landraces* which farmers have adapted over centuries of selection and use to meet dynamic and changing needs (Worede 1993). Ethiopian farmers are also instrumental in conserving germplasm as they control the bulk of the country's genetic resources. Peasant farmers retain some seed stock for security unless circumstances dictate otherwise. Even when forced to temporarily leave their farms because of severe drought or other threats such as war, farmers have often stored small quantities of seed stock for later use. They usually employed various containers such as clay pots or rock hewn mortars (in Northern Shewa, for example) that were sealed and buried in an inverted position in a secured place on the farm and in underground pits (Figure 6.1)

In addition to household storage, farmers in various regions of the country have well-established systems to ensure the sustenance of seed supply, and they often operate in networks. One of the principal networks is that of the exchange of seed in local markets. Farmers exchange crop types representing a wide range of adaptation to diverse environments. In this way, farmers benefit from a wider choice of planting material to suit a particular set of agro-climatic conditions. Seed that is not exchanged or consumed can be saved for a more appropriate planting season. In some of the more developed regions of Ethiopia, such as the central highlands, this practice is becoming less and less common with the availability of new, improved cultivars. In most of the drought-prone areas, particularly in the northern Shewa and Wello regions, farmers still depend largely on the above-mentioned traditional system of ensuring sustained supply of adaptable planting material (Hailu Getu 1991, personal communication).

Ethiopian farmers have been instrumental in creating, maintaining, and promoting crop genetic diversity through a series of other longstanding activities. On many peasant farms, cultivated crops often intercross with their wild or weedy relatives growing in the same field or in nearby fields. This results in new genetic combinations that farmers can use to meet agro-ecological realities. Similarly, small-scale farmers in various regions of Ethiopia quite frequently practice intercropping, or even grow their crops in mixtures, to stabilize their crop production, especially under adverse growing conditions. Genetic introgression within these mixes leads to rapid diversification among the included species. This has apparently happened in the *Brassicas* where new and different characteristics occur in *Brassica carinata* (Ethiopian mustard) and *Brassica nigra* (black mustard) on farms where such crops are grown in mixtures (Worede 1987).

---

* Landraces are crop plant populations that have not been bred as varieties but have been adapted through years of natural and artificial selection to the conditions under which they are cultivated. They could also be referred to as "folkseeds" to reflect the role of local communities in selection and innovation. See also the final consensus Report of the Keystone International Dialogue Series on Plant Genetic Resources, Madras Plenary Session, Second Plenary Session, 29 Jan. – 2 Feb. 1990, Madras, India.

Figure 6.1   An underground seed storage pit used by a farm household in Ejere, Central Shewa.

With coffee (*Coffea arabica* L.), farmers often plant populations of local types on small areas adjacent to the more uniform Coffee Berry Disease (CBD)-resistant lines which are distributed by the Coffee Improvement Project in Ethiopia. The coffee project is a source of tremendous support for EBI efforts to maintain coffee germplasm in the field because it is difficult to safely store the crop on a long-term basis as seed. EBI has also benefited from the knowledge and skills of farming families collaborating in genetic resources activities, especially in collecting germplasm and identifying useful plant material including plants with potential for industrial and medicinal uses. This has already contributed to the availability of information on the country's crop germplasm resources that these farmers have developed and maintained for many generations.

The threat of genetic erosion

The broad range of genetic diversity existing in Ethiopia, particularly the primitive and wild genepools, is presently subject to serious genetic erosion and irreversible losses. This threat results from the interaction of several factors and is progressing at an alarming rate. The most crucial factors include the displacement of indigenous landraces by new, genetically uniform crop cultivars, changes and development in agriculture or land use, destruction of habitats and ecosystems, and drought.

The drought that prevailed in the regions of Wello, parts of Shewa and Northern Ethiopia, has directly or indirectly caused considerable genetic erosion, and at times has even resulted in massive destruction of both animals and plants. The famine that persisted in some parts of the country has forced farmers to eat their own seed in order to survive or to sell seed as a food commodity. This has often resulted in massive displacement of native seed stock (mostly sorghum, wheat, and maize) by exotic seeds provided by relief agencies in the form of food grains. To counter losses in genetic diversity, PGRC/E has launched rescue operations during this period (1987–1988), including a strategic seed reserve program, in areas subject to recurring drought, as shown in Figure 6.2 (Worede 1991).

The extent to which the displacement of native seeds by exotic or improved materials occurs in Ethiopia has not been fully documented. Rates of displacement vary depending on regions and crops. In many cases, farmers still plant both native and exotic types interchangeably or alongside each other, at times in mixtures, depending on their particular need, market demand, or other prevailing factors.

In general, native barley and durum wheat are among the crops most threatened by new varieties and/or by other crop species such as teff and bread wheat, which are expanding within the cereal growing highlands of the Shewa, Arsi, and Bale regions, largely because of greater market demand. Similarly, in the central highlands, including the northern Shewa and Gojam regions, introduced varieties of oats are expanding rapidly, often replacing a wide range of cereals, legumes, and pulses grown in these areas.

With sorghum and millet, exotic varieties do not pose any immediate threat because expansion of such materials is at present somewhat restricted. In the case of sorghum, however, genetic erosion is progressing on account of extensive selection and breeding of the native populations. The Ethiopian Sorghum Improvement Project (ESIP) has been doing extensive mass selection on sorghum and millet and, in some cases, selecting single lines or

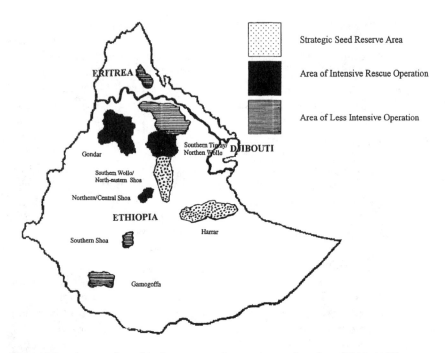

Figure 6.2　Areas where landrace operations were undertaken (1989–1988).

cultivars to develop elite materials with improved yield and/or disease (smut) and pest (stalk borer) resistance. The distribution of these materials results in a gradual displacement of the original farmers' seed stock, especially in the regions of Wello and South East Shewa. A similar situation exists with the various pulses, legumes, and oil crops grown in the country, where the bulk of the material utilized in breeding programs is represented by indigenous landrace populations. For crops such as sorghum, millet, and pulses, for which there is no immediate threat of genetic erosion, there still exists a danger of their massive displacement in the future by the expansion of other crops with better market values (e.g., maize, teff), monocopping, and shifts in cropping patterns that favor early maturing varieties.

Need for research to conserve and enhance in-field diversity

In the context of peasant farms, *in situ* conservation is defined as the maintenance of traditional cultivars or landraces in the surroundings to which they have adapted, or in the farming systems where they have acquired their distinctive characteristics. Duvick refers to *in situ* conservation of landraces as "evolutionary conservation" (1991, personal communication). As such, *in situ* conservation will help sustain the evolutionary systems that are responsible for generating genetic variability and provide, therefore, a valuable option for conserving crop diversity (Worede 1991). Maintaining this dynamic process is especially significant in regions of the country subject to drought and other stresses, because it is under such environmental extremes that variations useful for stress-resistance breeding are generated. In the case of diseases or pests, this would allow for ongoing host-parasite co-evolution. The ability of landraces to survive under stress is conditioned by a broad genetic base, inherent to landrace populations. In contrast, the more uniform, new, or improved cultivars — despite their high yield potential — are less stable than landraces that have been maintained under adverse growing conditions.

Landrace evaluation and enhancement programs will certainly be needed to promote more extensive utilization of germplasm resources that are already adapted to drought-prone regions of Ethiopia. Under such extreme environments, locally adapted landraces would provide suitable base materials for institutional crop improvement programs. There is, therefore, a need to maintain landraces growing under natural conditions in a dynamic state. In Ethiopia, maintaining landraces is probably best achieved through farm- or community-based conservation programs.

The work described above began in Ethiopia in 1988, when PGRC/E with support from USC/Canada, implemented the SoS/E Program. The program continues to represent a participatory, dynamic, farmer-based approach to landrace conservation, enhancement, and utilization. The activities of SoS/E are linked to the more formal off-farm conservation activities at the national gene bank (EBI). The work is carried out on small-scale peasant farms in collaboration with local farmers, scientists, and local extension agents. The program is comprised of two major types of farm-based conservation activities: conservation and enhancement of native seed stock (landraces); and maintenance of indigenous landrace selections (elite materials) on selected farms (Worede 1992). The salient features of these and other related activities are described below.

Landrace conservation and enhancement on the farm

Genetic resource conservation and enhancement activities involving farmers, scientists, and local extensionists began in 1988 and are now expanding within a network of selected farms at strategic sites in areas where the native seeds are still widely grown and where stresses such as recurrent drought, disease, and pest epidemics prevail. SoS/E designed its conservation measures primarily to maintain in-field crop diversity by protecting major cultivars from disappearing, and to improve the genetic performance of diverse landraces. Targeted crops include sorghum, various pulses, and locally adapted maize. Materials collected (or rescued) during the drought period (1987–1988) are included in the program. Farm families participating in the initiative were initially selected and organized through their respective farmers' cooperatives.

Landraces are maintained on each peasant farm (Figure 6.3) following the traditional practices of selection, production (including weed management), storage, and utilization. Field sites vary each season in conjunction with the traditional crop rotation patterns. The plot size and seed rates for each crop are determined by the farmers, depending on farmers' needs, availability of seed and labor, method of seeding, and soil type. In managing and maintaining the *in situ* plots according to farmer practices, the program seeks to to optimize *in situ* conservation, based on the rationale that farmer practices provide a viable approach to long-term conservation.

Figure 6.3   A farmer-based sorghum landrace *in situ* conservation plot at Terefo, Southern Wello (1995).

Identification and establishment of strategic *in situ* "pockets" over a network of locations is another major component of the project. At present, this is limited to identifying strategic sites in locations where the targeted landraces are grown, spreading across a range of agro-ecological niches within the project area. Not all identified spots are necessarily planted to *in situ* crop materials at one time, or with the same populations of a crop species every cropping season (Table 6.1).

Studies are underway to document and build on the existing knowledge and practices relating to landrace production and management on selected farms. Additional scientific inputs are needed in the areas of socioeconomics, ethnobotany, and population biology, focusing on the population structure and dynamics of the various landraces for more effective planning and management of *in situ* conservation strategies. EBI is currently conducting research in these areas, on a multidisciplinary basis, involving farmers who play a key role in providing information on studies related to the ethnobotanical and socioeconomic aspects, as well as on the general biology of

Table 6.1 Distribution of Sorghum Landrace Types *in situ*: Conservation Farms across a Range of Agro-Ecological Niches in Wello Region*

Location	1993–1994	1994–1995	1995–1996
Laygnaw Attaye	2	7	—
Merewa Adere	2	2	—
Hora Dildye	1	—	—
Kemisse	1	3	1
Terefo	2	5	23
Fontanina	1	5	—
Chefe Mesendi	—	3	9
Kedida Albuko 02	—	5	—
Kedida 02	2	12	5
Loga Haik	2	1	—
Batti	2	2	1
Kobbo	2	2	8
Ashenge	2	—	—
Korem	1	—	—
Bizet	1	—	—
Agulla	1	—	—

* Seed not planted the following season on a given location is stored safely for another planting season, while the same landrace is planted elsewhere across a range of agro-ecological locations.

their surroundings. Women farmers, in particular, are encouraged to participate because, despite their crucial role in providing valuable information, they are often dominated by their male counterparts. Much of this work is in the initial stages, with greater emphasis thus far on the multiplication and distribution of elite landraces and the limited resources available.

In addition to EBI activities, farmers collaborating in the project practice various forms of stratified and mass selection and multiply their landraces (mainly sorghum and local maize) separately for production. Seeds of selected plants are bulked to form a slightly improved population which is included in plantings to increase seed supply and for continued selection (Figure 6.4). An appreciable amount of improvement in crop yield has been observed among the selected materials that are produced following the traditional systems. Yields of the sorghum landraces and locally adapted maize which have been jointly selected by farmers and EBI scientists have exceeded the yields of the original landrace seeds, with no additional input (Ataro et al. 1994). Farmer-selected types are expanding into other areas of the Shewa and Wello regions where frequent crop failures have occurred due to prevailing droughts. To date, 3,102 farmers are using the sorghum varieties and 2,999 farmers are planting the maize, and the number may grow to 18,000 by 1999 — with 6,000 new farmers receiving seed each year. Seed provided to these farmers by SoS/E (usually

Figure 6.4   Farmer selection of sorghum landraces multiplied at Terefo, Southern Wello.

20 kg) is retrieved at the end of each harvest and stored for redistribution the following season.

Representative samples are drawn from both the elite and original farmers' seed and stored at the national gene bank, for further evaluations of various characteristics useful in local and institutional crop improvement. Gene bank activities provide an opportunity for transferring genes that control characters of interest (e.g., disease/pest resistance, drought tolerance, and high lysine in sorghum) from existing selections or from external sources to enhance the elite populations.

Currently, some 500 peasant farmers are paid on a contractual basis for conserving materials that are likely to disappear or be abandoned but might have potential value, and for multiplying elite landrace materials for distribution to local farmers in the region. Payment is determined on the basis of the additional inputs (labor and various costs) incurred by participating farmers. Farmers work closely with local SoS/E field staff to monitor on-farm conservation activities.

Maintenance of elite indigenous landrace selections on peasant farms

Another aspect of the program deals with restoring landraces in regions where farmers had once planted landraces extensively, but which are now dominated by introduced or improved (high external input) varieties. In the region of Ada in Central Shewa, for example, the indigenous durum wheat has nearly disappeared because of displacement by introduced bread and durum wheat varieties. In this area, farmers (primarily women) traditionally used the local durum wheat to make porridge, *enjera*, unleavened bread, and homemade beer, which they sell or exchange at local markets. Farmers rarely use bread wheat for household consumption; rather, they sell it as a commercial crop in urban areas.

SoS/E has been active in promoting the conservation, enhancement, and utilization of indigenous durum wheat in Ada and other areas of Central Shewa. Elite durum wheat landraces (composites of three or more genetic lines) are developed at the Debre Zeit Agricultural Research Centre of Alemaya Agricultural University (DZARC/AUA) and provided to SoS/E (Tesemma 1996). These composites are further selected and multiplied jointly with small-scale peasant farmers. The project currently includes nearly 4,000 farms in Central and Eastern Shewa. Landrace composites were developed from plant populations subjected to selection based on performance in yield tests under different conditions of environmental stress. Genetic lines (agrotypes) are bulked for further selection, multiplication, and distribution to farmers (Tesemma 1987). The program utilizes EBI wheat collections as well as materials collected during the last 12 years by the durum wheat breeding team of DZARC/AUA in close collaboration with EBI.

Considerable progress has been made in yield improvement over the past 8 years through further selection from some of the most productive landrace materials. Initially, participating farmers received 68 composites, 50 of which were selected for multiplication and evaluation at the various sites (Table 6.2). Since the 1994–1995 cropping season, eight composites most

Table 6.2   AUA-SoS/E Durum Wheat Landrace (Elite Agro-Types) Multiplication and Testing Sites (Farms) during the 1994–1995 Crop Season, Central Shewa Region

District	Locality	No. of Composites	On-Farm Conditions
Ada	Dirre	9	Low rainfall zone/recurrent drought
Ada	Godino	10	Water logging
Loumae	Ejere	21	Frost
Ambo	Ambo Amaro	10	Water logging

preferred by the farmers have been under production at various locations on 4,000 farms in the above-mentioned regions. Farmer demand for landrace seeds has been escalating at an impressive rate. As frequently observed during field visits, the elite seeds are also finding their way to farms outside of project premises, most likely through informal seed exchange or diffusion of seed at local markets.

In a preliminary comparative yield assessment conducted in the project area over the past 3 years the elite durum landrace selections (composites) generally out-performed their high input counterparts, which are represented by improved, high yielding varieties (HYV). The yield performance of these elite materials on the peasant farms was astoundingly high, compared to both the original farmers' seed or the most predominant HYV (Boohie) during the 1994–1995 crop season (Tesemma 1996). On one peasant farm in Central Shewa, for example, the highest-yielding composite out-performed the local landrace by 37% and the HYV by 40%, while no important difference in yield was observed between the latter two (Table 6.3). Similarly, in Deka Bora, Central Shewa, the top yielding composite out-yielded the HYV and the local cultivar by 127 and 80%, respectively (Table 6.4). In both localities, the yield performance of the composites exceeded those of the HYV and farmers' varieties by an average of 25 and 10%, respectively.

Table 6.3   Yield Estimate (kg/ha) of Elite Farmers' Varieties, Composites Sampled at One Locality in Godino, 1994–1995 Cropping Season

Composite No.	Sample						Average Yield
	1	2	3	4	5	6	
3	1000	1120	1750	1750	1750	2250	1603.33
5	1750	1500	1620	2500	1500	3370	2040.00
7	1820	2120	1320	1750	1250	1370	1605.00
9	1370	1250	1370	1370	1370	1250	1330.00
12	1250	1500	1500	1250	1000	1620	1353.33
16	1100	1100	1250	1500	1250	1620	1303.33
20	1600	1500	1250	1380	1250	1250	1371.67
24	2250	1125	1500	1750	1370	1250	1540.83
30	1060	1350	2120	1870	1620	1370	1565.00
31	2250	2120	1320	1380	1820	2000	1815.00
32	1000	1250	1500	1500	1620	2050	1486.67
Arendeto*	1000	1250	1620	1500	1620	2250	1540.00
Boohe*	1500	1500	1600	1250	1750	1120	1453.33
Mean							1539.04
S.E.							192.80
LSD(0.50)							544.50
CV							12.52

* = Farmer's variety

**Improved, high input variety (IHYV)

Date planted = July 10–15, 1994

Date harvested = December 21, 1994

Previous crop = Vetch and chickpeas

The HYV was poorly adapted to the adverse growing conditions that prevailed during the growing season, especially to periods of drought and frost that occurred at the heading stage. This is often the case with the high input exotic varieties which were developed from a narrow genetic base for broad adaptation. Frequently, modern varieties fail to niche into the set of

Table 6.4 Yield Estimate (kg/ha) of Elite Farmers' Varieties, (Composites) Sampled at One Location in Deka Bora, 1994–1995 Cropping Season

Composite No.	Sample						Average Yield
	1	2	3	4	5	6	
53	750	1750	1000	1380	1180	1380	1240.00
54	880	1870	1500	1500	1250	1000	1333.33
55	750	1120	1250	1120	1750	1000	1165.00
57	1000	880	1120	1380	880	1120	1063.33
58	1000	1120	1000	1250	1000	880	1041.67
59	1250	1120	1750	750	1370	2750	1498.33
60	1750	1500	1630	2500	1500	3000	1980.00
61	1870	2120	1370	1750	1250	1370	1621.67
31	1370	1250	1370	1370	1370	1250	1330.00
Arendeto*	850	1500	750	1000	1250	1250	1100.00
Boohe**	1250	1000	750	1000	490	750	873.33
Mean							1295.16
S.E.							207.17
LSD(0.05)							587.14
CV							15.91

* Farmer's variety

**= improved, high input variety (IHYV)

Date planted = September 1, 1994

Date harvested = March 3, 1995

Previous crop = Faba bean

environmental conditions that are specific to the small-scale peasant farms. In addition to agro-climatic conditions, cultural practices such as the rate and timing of fertilizer applications, seed bed preparation, planting method, weed management, and pest and disease control are important factors impacting the potential of the HYVs grown in these areas. Farmers will continue to multiply and use composites that are best suited to their conditions along with other landraces and improved materials provided by the national breeding programs. This will allow farmers to critically evaluate the source of planting material, which now consists largely of relatively poorly adapted HYVs distributed to farms throughout the region. It also encourages the farmer to make continued use of landraces, and ensures effective utilization of superior germplasm, avoiding the threat of losing unexplored germplasm represented by the indigenous population.

The activities described above represent a form of *in situ* conservation of indigenous durum wheat selections (composites) designed to promote productivity (higher yield) while retaining an appreciable amount of variability that exsists in the landraces from which these materials were developed. The elite landrace selections provide a dependable source of planting material with the potential to out-perform the improved, exotic seed that often fails to meet farmers' diverse needs and requirements. The program is working toward adding new entries to the pool of elite landrace selections (mostly composites) to meet a greater diversity of needs and the growing demand for composite seeds.

In conjunction with the development of elite landrace seeds, the original farmers' seed stock is conserved *in situ*, following the same strategies described for sorghum and maize above, to represent a backup system for the rapidly expanding elite durum landrace selections. The *in situ* plots are maintained and further developed by EBI within a network of farms representing the various agro-ecological niches in the SoS/E project area, as part of EBI's wheat *in situ* conservation activity now expanding to other areas in the Northern Shewa, Gonder, and Tigray regions (Table 6.5). EBI scientists also collect representative landrace samples for storage at the national gene bank and evaluate and characterize these materials for use in local crop improvement programs.

Table 6.5   A Network of Durum Wheat Landrace *in situ* Conservation Plots in Central Shewa

Agro-Ecoregion	Locality	Elevation (m)	Farm Size
Ejere	Tiriti	2000	0.3 ha
	Illa bela	2200	0.25 ha
	Tulu Iola	1800	0.2 ha
	Ejere	2400	0.3 ha
Godino	Ganda gorba	1800	0.3 ha
	Ganda gorba	1600	0.2 ha
Dire	Chelleba	1600	0.2 ha

Source: Adaa agro-ecoregion on-farm activity of 1996 crop season, progress report (EBI).

Building on existing activities: adding new dimensions

The pioneer work of SoS/E continues with new initiatives emerging as the program probes the immense possibilities of supporting community-based genetic resource activities, building on the knowledge and skills of peasant farmers. One such initiative is to enhance small storage units (underground pits, clay pots, etc.) that farmers traditionally use to store seeds for planting, particularly special seed selections maintained by women farmers. Improving storage units will help to preserve diversity more effectively, thereby complementing the more formal *ex situ* system. Maintaining seed stock in this way would also provide a backup system for *in situ* field plots, in case of crop failures, and thus provide a mechanism of ensuring the continuity of on-farm landrace conservation.

SoS/E is working to develop new strategies and approaches for landrace utilization, adding a new dimension to its yield enhancement efforts, including the promotion of elite landraces selected and enhanced on the basis of growing urban consumption needs. Urban demand for landraces would provide market incentives for farmers to produce indigenous seeds beyond the subsistence level. Through collaboration with peasant farmers, SoS/E scientists have already identified a few elite seeds (e.g., white and purple seeded durum wheats) with the potential for use in the food industry, particularly for pasta and pastries, which at present depends largely on imported food grain. SoS/E is multiplying these seed types for both local and urban consumption.

Long-term stability of food crop production may be ensured by maintaining a wide array of landrace materials or cultivars, which farmers traditionally maintain to adjust to new, changing conditions, including market demand. Complementing this with improved farming practices (e.g., crop rotation, soil and water management, etc.) is, however, crucial to sustained crop yields that SoS/E is presently undertaking as part of a comprehensive program to improve overall farm productivity. The success of the Ethiopian program has led to the emergence of new initiatives that are now building on existing farmer-based landrace conservation, enhancement, and utilization. The Global Environment Facility (GEF) has allocated approximately US$2.5 million to support a comprehensive program for an initial 3-year period. Among other activities, the GEF project will build a series of community gene banks (CGB) to maintain local germplasm for crop improvement and a seed reserve system for emergency use.

The perspectives

The farmer-based genetic resource program in Ethiopia has made major strides toward building national food resources at a time when Ethiopia is seriously threatened by famine and irreversible losses of its crop diversity. Traditionally, peasant farmers have maintained a sufficient amount of field diversity to sustain productivity and diversify dietary needs. Such diversity has allowed farmers to maximize output under adverse farming conditions and environments. Traditional varieties or landraces are well adapted to these environments and produce stable yields over changing seasons. Maintaining a sustainable balance between conservation and production has been a major challenge for many African countries that have adopted Western agricultural models to increase food production. Such models include the expansion of a limited number of major crops for cash-crop production and the increased application of inputs that are often difficult for small-scale farmers to obtain. Rather than capitalizing on natural resistance, the application of chemical inputs to fight pests and diseases has become more routine; crop rotation systems may be replaced by successive planting of a single crop, which often requires fertilizer inputs.

The Ethiopian initiative is, therefore, a timely venture that seeks to improve agricultural production without displacing existing cropping systems and diversity. The main objective is to help peasant farmers retain diversity while improving productivity and to maintain the freedom of choice with regard to planting material. The program's success is largely attributed to the significant number of farmers in Ethiopia who are now benefiting from the use of landraces that they themselves have selected and multiplied.

Future projections*

The involvement of farmers in the conservation and utilization of Ethiopia's germplasm resources will be strengthened and expanded to cover a broad range of agro-ecological conditions and strategic sites. The process is a challenging one, demanding a comprehensive knowledge of the country's vast resources and the diverse farming and land-use systems that have maintained these resources. Rationale for developing a more comprehensive network for *in situ* conservation of landraces is already fairly well established in Ethiopia, as in other centers of diversity (Altieri and Merrick 1987; Brush 1991). *In situ* conservation is recognized as a participatory and dynamic approach, involving farmers and their long established skills and knowledge of landraces, and allowing continual evolution and generation of useful germplasm. It is relatively inexpensive given the amount of potentially useful material that is preserved. As a complement to *ex situ* measures, *in situ* conservation represents a viable and vital component of Ethiopia's long-term strategy to effectively conserve and utilize genetic resources.

Planning and implementing an effective *in situ* conservation program will require prioritization of crops and landraces. In Ethiopia, landraces of the crops targeted for an *in situ* conservation program may not be adequately or safely maintained in the few areas where sites are presently being established. With each crop species farmers spread their risk across time, space, and the diversity of the material they grow. This occurs at the farm-household, community, and regional levels. Farmers' exchange or diffusion of both material and information about their seed may account for the wide range of adaptability as well as the plasticity inherent in these materials. It is, therefore, essential to plan a correspondingly wide network of *in situ* conservation sites, taking all of these factors into consideration, supported by more extensive research relating to the genetic, ecological, and social dynamics of landraces.

In the long run, *in situ* conservation work will be expanded to conserve wild plant species in their natural habitats. Plans are already under way to conserve *in situ* wild relatives of cultivated crops and wild plant species of potential value at strategic sites in areas where diversity exists. This may be undertaken as part of a community land management program, in areas surrounding farms, where such materials still exist but are progressively diminishing due to changes in land use such as increased grazing or agricultrual expansion. A community program may extend beyond crop species to include several trees, shrubs, and grasses that grow wild and are traditionally used by communities for food, medicine, and fuel. As part of the

* More details may be found in relevant EBI Activity Reports and Project Documents.

national coffee conservation program, for example, a special effort is being made to conserve the semi-cultivated coffee (*Coffea arabica*) on peasant farms in three regions of Ethiopia (Kefa, Illubabor, and parts of Wellega), in areas where the forest coffee occurred spontaneously. These efforts will complement the field collection now being maintained at Chochie, Kefa (PGRC/E 1992).

Linking landrace conservation to utilization

The value of landraces to farmers in a developing country like Ethiopia lies in their utility as a dependable source of planting and breeding material. It is important, therefore, that locally adapted or enhanced seeds are multiplied for distribution to farmers whose production requirements have not been adequately met by modern high input cultivars. Unless the landrace conservation activity is oriented toward supporting sustainable production, local farmers may see little value in conserving landraces.

One of the strategies of the Ethiopian on-farm landrace conservation and improvement program is to acquire an understanding of the distinct challenges in potential program areas, prior to initiating any activities. Information is collected on farmers' needs and objectives, agro-ecological requirements, distribution of existing diversity within and among crops species over locations, the level of the diffusion of improved materials in the area, and farmers' conservation strategies (Feyissa 1996). Based on this information, mechanisms are designed for linking conservation to production in order to address farmers' needs and objectives. Improvement of the potential of landraces, development of markets for farmers' products, establishment of community-based seed networking, and integration of conservation strategies at all levels are the priority areas considered necessary to sustain a system (Figure 6.5). Existing networks such as community-based seed production or marketing and distribution systems will most likely be the best way to address the needs of local farmers. Through this approach, farmers will be able to control the choice of crop types and cultivars targeted in a conservation program, and they will have ready access to planting material adapted to local growing conditions. Farmers will also be in a position to critically evaluate the relative merits of a wide range of cultivars, thereby limiting undue expansion of exotic cultivars that are costly and poorly adapted.

Furthermore, the community seed bank is a low cost and low technology system that will be owned and managed by local communities as part of existing community service cooperatives (Worede 1997). The community seed bank is comprised of two major components: a seed store and a germplasm repository for local crop improvement, complementary to the gene bank at EBI. The seed store represents a seed reserve system, consisting largely of landrace materials developed and multiplied contractually by farmers. The store provides a backup to the local (informal) market networks, where

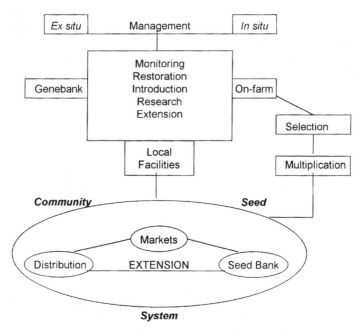

Figure 6.5   Conservation-utilization link. (Source: Feyissa 1996.)

farmers traditionally exchange seeds and information. The seed reserve that the community seed banks maintain becomes crucial to ensuring a sustained supply of adapted seeds, channeled through the informal market system, thereby averting the potential loss of genetic diversity. Finally, traditional storage units, such as clay pots, rock hewn mortars, and underground pits, form an integral part of local seed storage systems. Improved versions of these small units could be established within a network of farm-households to complement *in situ* networks (Worede 1997).

Concluding remarks

Genetic resource conservation represents a major national effort in Ethiopia, beginning with the establishment of PGRC/E in 1976. This work represents a unique opportunity to conserve landraces in a dynamic, participatory way, involving farmers who manage the bulk of the country's indigenous crop genetic resources, and in fact practice *in situ* conservation as a part of their traditional management strategies. The program is working to provide farmers with a wider choice of planting material, thereby encouraging sustained supply and use of locally adapted landraces, especially in marginal or stress environments, in which such materials generally perform more competitively than their high input counterparts. Despite astounding progress, major gaps in the knowledge of and approaches to crop *in situ* conservation persist in Ethiopia, considering the complexity of the farming systems and agro-ecological conditions under which the various crop species and their landraces are managed. There is also an outstanding need to support the farmer-based approach to *in situ* conservation by more extensive research on the genetic, ecological, and social dynamics of landraces.

References

Adare, A. and B. Tsegaye. 1994. *Survey on Relative Performance of Landraces for the 1993/94 Cropping Season*. Consultancy Report, Seeds of Survival/Ethiopia, Unitarian Service Committee of Canada.

Altieri, M.A. and L.C. Merrick. 1987. *In situ* conservation of crop genetic resources through maintenance of traditional farming systems, *Economic Botany* 41:86–96.

Brush, S.B. 1991. A farmer-based approach to conserving crop germplasm, *Economic Botany* 45:153–165.

Feyissa, R. 1996. Integrated conservation and utilization of agricultural crop diversity: a strategic option for linking conservation to utilization. (Submitted for publication, Biological Society of Ethiopia, Addis Ababa, Ethiopia).

Tesemma, T. 1987. Improvement of indigenous wheat landraces in Ethiopia. In *Proceedings of the International Symposium on Conservation and Utilization of Ethiopian Germplasm*, J.M.M. Engles (ed.). 13–16 October 1986.

Tesemma, T. 1996. Low agricultural inputs: Opportunities on farmer based approach to landrace conservation enhancement and utilization. Proc., Christian Relief and Development Association (CRDA) Workshop on Agricultural Extension, 19 March 1996, Addis Ababa, Ethiopia.

Worede, M. 1987. Conservation and utilization of annual oil seed genetic resources in Ethiopia. In *Oil Crop Proceedings*, A. Omran (ed.). Third Oil Crops Network Workshop, Institute of Agricultural Research and International Development Research Center, 6–10 October 1986, Addis Ababa, Ethiopia.

Worede, M. 1991. Crop genetic resources conservation and utilization: an Ethiopian perspective, *Science in Africa: Achievements and Prospects*. Proceedings of the Symposium of the American Association for Advancement of Science, 15 February 1991, Washington, D.C.

Worede, M. 1992. Ethiopia: A gene bank working with farmers. In *Growing Diversity*, D. Cooper et al. (eds.). London: Intermediate Technology Publications.

Worede, M. 1993. The role of Ethiopian farmers in the conservation and utilization of crop genetic resources. In *International Crop Science Society of America*, D.R. Buxton et al. (eds.). Madison, WI: Crop Science Society of America.

Worede, M. 1997. Ethiopian *in situ* conservation. In *Plant Genetic Conservation: The in situ Approach*, N. Maxted, B.V. Ford-Lloyd, and J.G. Hawkes. (eds.). London: Chapman & Hall.

Section IV

Policy and institutional issues

chapter seven

Optimal genetic resource conservation: in situ *and* ex situ

Timothy Swanson and Timo Goeschl

Introduction

A diversity of crop genetic resources has long been viewed as a means of increasing both global and local food security (Ehrlich et al. 1993) and as an important source of income security for farmers in the less developed regions of the world (Hazell and Norton 1986; Fabella 1989). It has been acknowledged as a mechanism of insurance against the risks that characterize agricultural activity. Here we focus on its importance for global food security, specifically as an input into agricultural research and development (R&D) sectors which concentrate on the solution of the continuously arising problems of pest and disease resistance in modern agriculture (WCMC 1996). Without continuous injections of "new" genetic material, these industries probably would be unable to resolve the recurrent problem of evolved resistance (Swanson 1996a). The question investigated here concerns the optimal means for ensuring continuing and permanent supplies of new germplasm for use in R&D for modern agriculture.

The retention of diversity by farmers to ensure their livelihoods is related to this issue. In the past the breadth of crop diversity retained by farmers in the developing world has been sufficient to maintain a substantial base of resources for agricultural R&D. There are two reasons why this base is threatened: (1) modern agriculture is a very different technology from traditional agriculture, relying upon homogenous, high yielding varieties and their linked inputs (chemicals, tools, and irrigation). As the frontier of the modern sector expands, homogeneity continues to replace traditional diversity (Swanson 1995); and (2) the development of economies more generally also hastens the end of the traditional farmer's reliance upon diversity. Crop diversification has been relied upon in the past as a means of securing individual insurance in isolated markets, but as markets expand and integrate the farmer substitutes other less-expensive methods of insurance based on financial and labor markets (Goeschl and Swanson 1996). Development of agriculture and rural economies has meant that the individual farmer no longer supplies the vast quantities of new crop genetic resources on which the agricultural industry has relied historically. Continuing development implies that this source will almost certainly disappear in the next few decades. Therefore, farmer-based conservation of germplasm is a related issue, because the decline of this practice is one of the fundamental reasons for concern about future supplies of germplasm.

What agent will replace the individual farmer as supplier of crop germplasm? There are two possibilities: the government and the private R&D sector itself (i.e., plant breeders). Both agents are already very active in their attempts to address this problem (WCMC 1996). Even if the private sector is able to identify and supply goods and services that society demands, however, it is not clear that it will do so to the socially optimal extent because the values concerned are both informational and long term in nature (Swanson 1996b). Agricultural R&D firms report that a single product cycle lasts around 7 years — the time it takes for a given crop variety to become economically nonviable due to the development of resistance (WCMC 1996). With a standard private sector discount rate of 10% or so, this would imply that the average R&D firm would have little interest in conserving germplasm to supply agriculture beyond one or two product cycles. Therefore, there is good reason to believe that the long-term supply of crop germplasm for agricultural R&D is a function that must be supplied by the government.

How should the government participate in the conservation of crop genetic resources? The range of instruments available to the government for this purpose are direct conservation (*"ex situ* policies"), direct intervention in farming practices (*"in situ* policies"), and indirect interventions across the broader agricultural industry (such as the reform of property right systems). This chapter concerns the question of how the government should allocate its efforts between the two former options: *ex situ* and *in situ* conservation policies.

In practice, specific positions have already been taken in regard to this question, as a substantial network of public gene banks has already been established. There is also an already existing commitment to this form of conservation, and commentators continue to call for investment in *ex situ* conservation strategies. For example, one analyst has called for the collection of all crop genetic resources related to rice production in *ex situ* storage facilities (Evenson and Gollin 1997). We would like to abstract from these pre-existing commitments and current conservation activities, however, and address the issues from first principles: What conservation strategies would be first best for conserving the values of crop genetic diversity? As these values are informational in nature, providing the options and insurance required to maintain modern agriculture, this chapter is a comparative assessment of how well two alternative instruments (*in situ* and *ex situ*) available to government function to conserve the values of genetic diversity.

In this chapter, we demonstrate that there are certain conservation functions regarding agricultural genetic resources that can only be fulfilled by means of *in situ* conservation methods. One important reason for this is that it is impossible to predict at any given point in time which genetic resources will be of value in the future. Optimal germplasm conservation concerns the solution of a continuously evolving problem, and hence *dynamic methods of conservation* are a necessary component of an optimal conservation strategy.

The first section describes the nature of the informational problems in agriculture; that is, it demonstrates that one of the basic inputs into the agricultural industry is the information required to resolve recurring problems of evolved resistance, and it also demonstrates how genetic diversity relates to the supply of both stocks and flows of such inputs. The second section then generates an abstract depiction of the informational problem in agriculture, showing how the management of agriculture depends upon the continuing control of various dynamic systems (biological and societal). The third section describes the optimal methods for conserving the information necessary for the maintenance of the agricultural system. The chapter concludes by arguing that there are necessary roles for both *ex situ* and *in situ* conservation, because both forms of conservation concern the use and management of different forms of information important for the continuance of agriculture.

Conservation technologies and the object of conservation

There are two basic technologies for storing crop genetic resources, *in situ* and *ex situ* (Orians et al. 1990). As in the biological literature (Frankel and Soulé 1981; Frankel et al. 1995), *ex situ* conservation of agricultural genetic material is usually conducted through the storage of specific samples in "seed banks": actual cold storage facilities for samples of crop germplasm (FAO 1996). The alternative, *in situ* conservation, consists of the continued cultivation of crop varieties by farmers, together with their practices of observation, selection, and use. Obviously, the two approaches are not two means of accomplishing the same thing. *Ex situ* conservation strategies are static in nature: they attempt to freeze the existing set of germplasm for later use. *In situ* conservation strategies are dynamic in nature: they allow the germplasm currently in use to evolve and alter. After any significant period of time, two originally identical sets of genetic material conserved via the two different strategies will differ substantially since they have been exposed to very different biological environments, and they capture different values as a result.

Ex situ conservation is highly developed with more than 6 million accessions stored worldwide, although there is a slight downward trend in the collection activity over the last few years (FAO 1996). National policies as well as private organizations are still committed to maintaining and expanding the current stocks, as there exists public recognition of the values associated with germplasm preservation. In the case of *in situ*, by contrast, there is empirical evidence that farmers are increasingly reluctant to provide this form of genetic maintenance, and that in the absence of counteracting policy (e.g., Brush et al. 1992), economic development will further reduce the farmers' incentives to do so (Goeschl and Swanson 1996). In short, *ex situ* conservation is the primary form of government intervention at present, while *in situ* conservation (previously provided by individual farmers as a matter of choice) continues to decline on account of the development of agriculture worldwide (Swanson 1995).

On first glance, the reliance of government on a strategy of *ex situ* conservation might seem to be justified by the relatively lower cost of managing germplasm in seed banks. Tolerable storage costs in combination with the immediate accessibility and relative safety of the material stored give reasonable arguments for favoring conservation in gene banks (Plucknett et al. 1987). Similarly there are no land use issues involved in *ex situ* conservation. The use of land in conservation implies the loss of production of high yielding varieties, and possibly the loss of least-cost production methods. For example, it has been argued that, by reducing the number of varieties in use, farmers are able to increase income by reason of productivity gains associated with specialization and comparative advantages in production (Fafchamps 1992). For these reasons, many of the initial considerations produce an unreservedly positive picture of *ex situ* preservation.

Inherent in this view, however, is a very specific perception of the problem which the selected conservation strategy is intended to solve. For such strategies to be successful, they have to come to terms with some fundamental laws governing agricultural activity and have to provide feasible solutions under this set of constraints beyond human influence. One of these laws is that agriculture is interdependent with several distinct processes that are in motion across time. The most immutable of these processes is that of biological evolution. By means of mutation and recombination there is continual nondeterministic change within the environment. This implies that the important agroeconomic traits of a given plant variety (mean yield, yield variance, pest resistance, water stress resistance) are defined in relation to a specific set of environmental conditions prevalent at a specific point in time. The values of individual traits and characteristics are therefore time-dependent (Evans 1993; Frankel et al. 1995). Given that we live in a biological world, the motion within this process will always be present and unavoidable. Economists view the nondeterministic change within the biological world as a flow of information, a perspective that raises fundamental questions concerning the optimal conservation and use of genetic resources under conditions of uncertainty.

What sort of conservation or use is required under conditions of uncertainty? There are three distinct issues: (1) how to ensure that the optimal quantity of information is acquired; (2) how to ensure that this information is then developed or processed in an optimal fashion; and (3) how to ensure that decision making then incorporates the processed information. In the context of modern agriculture, these issues concern the optimal supply of diverse germplasm to the plant breeding industry, for its use in the maintenance of a stable system of agriculture. How does diversity function as an input within this industry, and what forms might its supply take?

Research and development and the industrial use of information

Research and development (R&D), i.e., the process by which new ideas are developed for application to common problems, almost always results in a new solution concept. If successful, it will be embodied within some novel product and marketed. Economists have long analyzed the research and development process as one of information creation, application, and diffusion (Arrow 1962). The theoretical concept of the R&D process is usually presented as a production process, itself dependent upon the application of various factors of production (machinery, labor) for the development of useful ideas. These ideas form the information base subsequently applied by researchers for solving economic problems presented by society. "Innovations" are then the products that embody solution concepts applied to address society's economic problems. Certain industries by their nature expend substantial proportions of their total available resources on the R&D process. For example, the computer software, plant breeding, and pharmaceutical industries are all R&D intensive industries, with over 10% of their gross revenues invested in the development of solution concepts. In the plant breeding industry, an average of 18% of annual turnover is allocated to breeding and research activities (WCMC 1996).

In the pharmaceutical and agricultural industries, the object of the industrial R&D process is most closely linked to genetic resources in the search for solution concepts. Agriculture and medicine should be conceived of as living defense systems rather than static technologies. That is, these fields of human activity consist of continuing efforts to combat the erosion of human-erected defenses against a hostile biological world. In agriculture, we continue to maintain a system that attempts to keep at bay the constantly evolving pests and predators of our primary food crops. In medicine, we continue in our efforts to defend against the same as they impact upon human beings more directly. In both cases the defenses are neither absolute nor perpetual; they are constantly eroding under the forces of nature.

Biological diversity is an important input to R&D in these industries simply because it contains information that has been generated within the relevant context. It is not any biological diversity *per se* that is the most useful input into important human industries, but rather it is the information to be gained from the characteristics that have evolved within a living environment that is most likely to make a contribution. Biodiversity is useful to our industries because of the manner in which the existing set of life forms has been selected (within a living, contested system similar to our own), which provides us with an already vetted library of successful strategies.

Much of the R&D process in the agricultural (and pharmaceutical) industries has been focused on the screening of the strategies that are operational in nature, and their development for specific applications in the industrial context. In the attempts to generate new strategies for combating pests, one of the most important inputs to the process is the set of natural templates which nature provides. These naturally supplied "resistance strategies" constitute information themselves, and they are often useful in the development of specific forms of information for direct application within human industry. This is the reason that the agricultural industry is able to turn to nature for a supply of important inputs. The next section describes in more detail the precise nature of these inputs in the context of agricultural R&D.

Stocks and flows of information in agricultural R&D

R&D will always constitute an important part of the agricultural industry simply because the biological world will continue to generate problems that must be solved. Much of the R&D concerned with the problems generated by the biological world are dealt with by the plant breeding sector. A recent survey found that R&D in this sector is becoming increasingly focused on the problem of pest resistance. Approximately 45% of new germplasm material used for breeding purposes now goes into the development of pest and pathogen resistance in crops (WCMC 1996).

Biodiversity operates as an input into this R&D process, both as a stock and as a flow. The screening of landraces already in use in traditional farming practices is an example of the use of existing stocks of information. Often all that is required for the industrial application of the stock of information accumulated within a landrace is the transfer of this information into the modern sector. In this use of landraces, the local community has accumulated the information as a stock within the plant varieties already in use. Both natural and human selection of plants may have occurred hundreds or even thousands of years. Thus a landrace may be conceptualized as an organism in which a series of beneficial selections have occurred in response to environmental changes (pests, climate stress) and farmer preferences. In other words, a landrace accumulates a stock of previously successful strategies. The screening of such landraces functions as an important part of the agricultural R&D process, providing source material for innovations in agriculture. The extent of the accumulated value of these selections within landraces is indicated by their relative value within the plant breeding industry. For example, Evenson (1996) estimates that a single landrace has had approximately 1000 times the impact on modern agriculture (in terms of increased production value) compared to a plant variety that has no history of use.*

---

* Evenson (1996) estimates that the value of the impact of a single rice variety landrace accession added by IRRI to improved varieties is approximately $86 million, while the impact of one added by a national agricultural research program is $33 million.

The information generated by nature always arrives initially as a flow. This happens, for example, whenever a particular type of pest invasion eliminates a large proportion of an existing crop. The survival of some individuals of any such crop variety is indicative of the presence of a strategy of resistance that is successful within the current environment. Analysis of these individuals might allow for the isolation of the trait or characteristic that confers this resistance, which might then be incorporated within modern agriculture. In this instance, the retention of a diversity of plant genetic resources is generating a flow of information for use within the R&D process which, after careful analysis, may result in successful innovations in terms of plant varieties. For a particular plant variety that has been subject to years of use and farmer-based selection, a flow of information may accumulate in the form of a stock of "resistance strategies"; the informational value of genetic resources, however, always originates as a flow.

Consider how the plant breeding industry makes use of naturally generated information in its R&D efforts through to its incorporation in a modern plant variety (Figure 7.1). Effective characteristics for new plant varieties develop naturally through the process of "natural selection": only varieties that are able to survive existing threats (pests and pathogens) remain. Since the set of threats is constantly changing, the natural environment continuously produces a flow of new information on the characteristics that are relatively fit under current environmental conditions. This naturally generated flow of information is labeled **Stage I** (Figure 7.1) and continues to flow from nature as long as some portion of land use is dedicated to the use of a wide range of plant varieties with relatively unknown genetic characteristics.

It is possible for these flows of information to be accumulated over time. "Traditional farmers" (**Stage II**) have survived by observing this naturally produced information and the consequent selection and use of beneficial traits and characteristics. In this way, traditional plant varieties are transformed into the accumulated history of the information that nature has generated and that farmers have observed and used. Their landraces constitute a stock of information on naturally generated resistance strategies that have been successful in varying environments over the years.

Stage III of the industry is where the modern plant breeding sector resides. In general the modern plant breeding industry has operated primarily through the collection and utilization of the set of landraces, and hence the stock of naturally produced information that is encapsulated within them. That is, modern agriculture has been based on the development of a particular crop variety that is an amalgam of some subset of the traditional varieties and its widespread use. The remaining stock of information derived from the landraces is then retained to deal with subsequently arising problems.

The R&D industry in agriculture has relied heavily upon the accumulated stocks of naturally generated information within landraces, but it is the supply of information generally that is essential and not the conservation

	Inputs/Outputs	Entity
STAGE III	*New Plant Variety*	"Plant Breeder" factors: Breeding methods Agronomic evaluation Germplasm (elite breeding lines, landraces, wild species)
STAGE II	*Landrace* (Natural Selection + Farmer Selection)	"Traditional Farmer" factors: Disproportionate use (farmer selection) Observation of natural selection Exchange
STAGE I	*Traits Selected* by Ongoing Biological Evolution	Land Use Decision + "Nature" factors: Natural Selection Reproduction/Recombination Diversity of plant genetic resources

Figure 7.1   The production of a new plant variety.

of any given stock. There is no value to maintaining a particular set of resources at least cost if these are not the resources that will be needed to solve a problem that arises at a future point in time. An optimal conservation strategy must conserve the mechanisms that supply solution concepts for time-dependent problems rather than a particular set of germplasm.

In the long run it will be essential to have mechanisms in place that will provide flows of naturally generated information. In addition to exploiting accumulated stocks of information within the existing landraces, it will remain just as important to have a flow of information to the modern plant breeder. For this reason, existing methods of *ex situ* conservation, which have focused on the conservation of the previously accumulated stocks of information, have to be matched with a mechanism for assuring an optimal future flow. It is predictable that private plant breeders would be focused on a short-term accumulated stocks approach to R&D, but the public sector should be concerned with supplying the long-term flow of information that is required to sustain agriculture once these stocks are exhausted.

To successfully implement a strategy aimed at providing a constant supply of information to "plant production," the nature of these dynamic processes has to be understood in greater detail. The next section of this chapter provides a more specific description of the problem that must be resolved through germplasm conservation. In the section that follows, we turn to the task of defining the important differences between the alternative forms of conservation, and examining which values each captures.

Defining the dynamics

In the previous section we demonstrated why information is the crucial concept for understanding the importance of genetic resources. We also explained how the information residing in the genetic makeup of crops is processed in the course of plant breeding, first by farmers, and then by industrial R&D. The purpose of this section is to discuss the sources of this information and to examine those that are most important in the context of genetic resource conservation.

Economic theory has identified two essential characteristics for systems to produce information: (1) motion through time; and (2) indeterminacy of the path of this motion. In the context of agricultural activity, the system best characterized by these components is nature, and the information generated by its motion is that of biological evolution. Although we know that evolution takes place constantly, decision makers cannot know which specific path it will take, as this depends on random genetic events discussed below. Decisions made in the present involve genuine uncertainty about potential outcomes because the precise nature of future conditions cannot be predicted. This is crucial as nature, i.e., the biological environment, determines the success of specific crops, such as yield potential, yield variability, and especially susceptibility to pests and pathogens.

Figure 7.2 depicts the interaction between human choices at different levels and the evolutionary process in nature and demonstrates some of the complexities involved in decision making in a dynamic setting. Individuals and governments attempt to make the correct choices regarding crop variety conservation, through a combination of crop use and other strategies. Governments may act directly to effect conservation, or create incentives for individuals to do so. The choices that are made result in a certain set of crop varieties in use, and another set of germplasm conserved through other efforts. These choices are "correct" in a time-dependent sense, i.e., relative to the existing information base, but these choices only result in a set of initial conditions that are then processed in nature. The biological system evolves continuously, introducing new mutations and generating information on valuable traits via natural selection. At the end of one "cycle" of these dynamic processes (e.g., one growing season), new information arrives to inform the next set of choices, and a different set of genetic resources results from that which was initially selected. The problem of optimal genetic resource conservation concerns the selection of the appropriate decision-making methodology to apply within such a dynamic environment.

Of course, biological evolution is not the only dynamic process with which agriculture is interdependent. Other essential processes include (1) technological change, which influences the productivity of a given set of crops produced in combination with complementary inputs and is capable

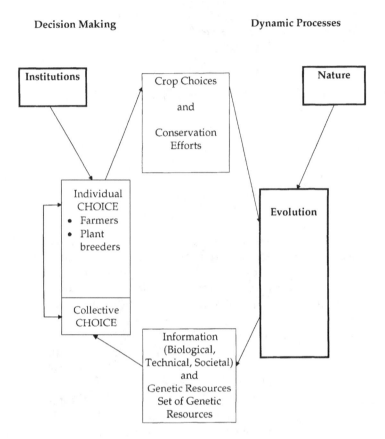

Figure 7.2   Dynamic processes influencing optimal conservation.

of designing new varieties; and (2) societal development, which is the source of the economic and non-economic valuation of genetic resources as well as of institutions which shape their exchange. There are strong systematic elements in their development, however. For instance, relative prices induce predictable responses, both in terms of production and technological development. As biological evolution is the most exemplary case of unsystematic change in the agricultural environment, our following discussion will mainly focus on this source of information.

In the remainder of this section we will outline the nature of the dynamics within the biological system. Since this process is continually in motion, any long-term decision-making process on the conservation of particular sets of germplasm must take this dynamism into consideration, i.e., it must make a decision on how to consider both current and future values that this process generates. For illustration, we will define a general agricultural production function which explicitly includes the biological environment as an argument in the agricultural production function Y_t.

Agricultural Production (time t):

$$Y_t = f(\Omega_t)^{\bar{\alpha}_{t,t_0}} \cdot \vec{p}_{t,t_0} \tag{7.1}$$

Agricultural output in time t, Y_t, is represented here as a function of the yield and variability of utilized crops. Information on these parameters is determined in the biological environment and is denoted by the time-indexed matrix Ω_t on which the choice of utilized crops depends. There is also a productivity parameter vector $\bar{\alpha}$, influencing the output valued according to the price vector $\bar{p}$. Note these two variables are subscripted t, t_0 to indicate that their values in t can be largely inferred from their previous state at t_0. Societal preferences are seen to alter by reason of changes in the price vector, while technical change is represented by movements across time in the productivity parameters. Nature provides information on the success of various characteristics in existing crop varieties, by reason of their apparent yields and variabilities in use.

The germplasm conservation problem concerns the selection of strategies that make optimal use of the existing stocks and future flows of information within the context of a changing environment. Before proceeding to the consideration of the optimal decision-making methodology, it is necessary to define with particularity the sources and forms of motion within the decision maker's biological environment.

Biological evolution

Nature introduces change into the environment via the depreciation of existing varieties in use and also by revealing the plant varieties that contain useful characteristics for the changing environment. The essence of the evolutionary process is one of change: mutation and recombination introduce new variants at each point in time. Since biological success (fitness) is a relative concept, the changing environment implies changes in the relative merits of individual varieties. This process of realignment may be broken down into two components: the depreciation of previously successful varieties and the revelation of newly successful characteristics.

Depreciation of existing varieties

Over a wide range of crops examined, yields in the most intensely cultivated varieties decline over time due to evolution in pests and diseases (Evans 1993). Investment into defensive technology is therefore necessary to maintain yield levels and to counter new biological threats to crops (Plucknett and Smith 1986). The basic problem of genetic vulnerability, however, cannot be eradicated unless the genetic composition of a crop d is continuously changed. This is especially true in the case of crops grown on a large scale such that epidemics can spread rapidly (Frankel and Soulé 1981).

The decline occurs not because the germplasm itself is depreciating but rather because the environment is shifting. As the environment changes, traits and characteristics that were effective under previous conditions are no longer effective under the altered ones. Although the process is embedded in environmental change, it is reflected in the observed yield results of particular varieties; specifically, the varieties which were most effective in previous environments demonstrate a tendency toward reduced effectiveness as the background shifts around them (National Academy of Sciences 1972).

The nature of this process of change is not necessarily continuous; for example, it might be the case that yields will continue relatively unaffected for several years and then suddenly collapse. It is more likely that the plant variety would have its resistance eroded by the aggregate effects of the loss of the productive capacities of a succession of useful characteristics. This would result in a relatively continuous rate of depreciation for a given variety. In any event, the available evidence indicates that over time, continued use of the same set of germplasm will inevitably result in problems of resistance. We will represent it as a continuous process of depreciation for any given variety d that causes expected yield to decrease over time at a rate δ.

Variety Depreciation (Loss of Resistance):

$$E(y_d) = \hat{y}_d \quad \text{and} \quad \dot{\hat{y}}_d = e^{-\delta t} \hat{y}_d \tag{7.1a}$$

Natural selection's "signals"

Changes in the background environment cause the previously "first-best" plant varieties to become less effective, but simultaneously these changes also reveal the traits and characteristics which are more effective in the new environment. This information is revealed through the process of natural selection, whereby relative fitness translates into increased representation in succeeding generations. Hence, investing land into a diversity of plant germplasm provides a flow of information as successive generations signal which plant varieties contain the traits and characteristics which are most successful under current environmental conditions.* Note that the continuing flow of this sort of information is dependent upon the use of lands for the planting of a wide variety of germplasm.

Therefore, over time evolutionary change (natural selection) is also capable of signaling the existence of characteristics for yield improvement opportunities. This capacity might be represented by a formal expression

---

* Natural selection may be thought of as operating in both the presence and the absence of human intervention. Natural selection can induce favorable changes across generations of untended plants, or it can help to reveal to the farmer desirable crop traits which had not been identifiable before. Owing to the observation and use of evolutionarily supplied information, traditional farmers have been able to develop highly productive and adapted landraces by deliberately selecting and breeding those plant characteristics revealed to be desirable by natural selection (Evans 1993; Harlan 1975).

demonstrating the prospects for a positive signal emanating from any given crop *d*.

Natural Selection's Signals:

$$\hat{\dot{y}}_d = \Delta \tag{7.1b}$$

where Δ is the realization of a random yield increase variable which is drawn in each period from an independently and identically distributed probability function $f(\Delta)$. The relative values of Δ will be comparatively small, however, and distributed closely to zero since they cannot compensate for biological depreciation of crops in extended use. To summarize these ideas, we have shown that the moving background represented by evolutionary change produces two outputs: (1) a systematically declining rate of production from the varieties already in use in modern agriculture, and (2) signals regarding those traits and characteristics which will be successful under the newly prevailing conditions. The receipt of these signals is dependent upon the investment of a quantity of lands in the widest possible sets of germplasm. These sites are then able to "reveal" information on successful characteristics in the new environment.

Conclusion: information relevant to genetic conservation

This section has identified nature as the important source from which information might flow to affect future decision making regarding the value of genetic resources. Other sources, such as technology and society, exist but do not exhibit such a large degree of randomness and indeterminacy as the biological environment. Genuine uncertainty in all three of these processes — especially in the case of nature — ensures, however, that information has a value as unpredictable developments occur and makes its reception and incorporation into decision making indispensable.

Defining conservation strategies: ex situ, in situ, *and optimal*

In situ and *ex situ* technologies of conservation represent fundamentally different approaches to problem solving. In this section we will define how these strategies differ in their solution of the conservation problem in the context of the dynamic decision-making environment outlined in the previous section. In essence, *in situ* conservation may be defined as an approach to decision making that is focused on the use of information arriving over time, whereas *ex situ* conservation is based on a commitment to a given set of germplasm at a given point in time. The relative values of the two approaches are dependent upon the expected value of the flow of information in the decision making context. When a flow of information across time is important, *in situ* conservation will afford additional values to those sup-plied via *ex situ* methods. The argument of this chapter has been that the essence of the problem of global food security in modern agriculture is the provision of a continuing flow of information, and hence *in situ* conservation provides important values for modern agriculture.

It will be necessary to evaluate each of the available approaches to conservation against a given societal objective. The objective here will be taken as the maximization of the objective function stated in Equation (7.1) across time; this gives the expression for maximum social welfare provided in Equation (7.1') below. In other words, we assess different strategies in relation to their contribution to maximum agricultural production over a given time horizon by — once again — producing output based in informa-tion Ω_t, technology set α_t, and evaluated at prices p_t. Future production is discounted at factor r.

Societal Objective in Agriculture:

$$\text{Max} \int_0^\infty e^{-rt} Y_t dt \equiv \text{Max} \int_0^\infty e^{-rt} \left(f(\Omega_t)^{\bar{\alpha}_t} \cdot \right) \bar{p}_t dt \tag{7.1'}$$

This is a very simplified version of the societal objective function regarding global agricultural production, which places maximum emphasis on maxi-mizing the values of global yields. This abstracts from other issues such as distribution, variability, and desirability, and focuses on the single issue: how should genetic resources be managed in order to provide for maximum global yields in agriculture?

In situ *conservation as a closed-loop strategy*

In situ conservation (as used here) implies the existence of a group of indi-viduals who continue to dedicate some amount of land use to a broad set of crop genetic resources under very flexible technologies. In the past, farm-ers in less developed countries effected precisely this as part of their opti-mizing behavior, using crop genetic resources as a hedge against financial risks, and farmers continue to do so in isolated settings. As markets mature, individuals have access to more efficient methods of hedging risk and replace *in situ* conservation with these other financial instruments (Goeschl and Swanson 1996). The objective of *in situ* conservation is to give some set of farmers an economic reason for continuing traditional farming practices. This means creating a system of incentives that will cause a group of farmers to act so as to simulate their previous behavior in a different economic environment, which is to maximize their risk-adjusted income by making use of the naturally sourced information available at every point in time when carrying out their cultivation decisions.

Possible incentive systems for *in situ* conservation might entail, for example, paying farmers to dedicate certain designated lands to the use of only those plant genetic resources acquired from the previous year's harvest. The challenge of successful *in situ* conservation in an altered economic environment is to realize simultaneously several objectives: (1) decentralized choice by individual farmers; (2) optimal conservation of genetic resources; and (3) minimum welfare loss for the farmers engaged in the conservation effort since these losses will invariably undermine farmers' commitment to this undertaking. Incentive structures have to be examined in this light whether they cause farmers — for example — to consider using plant genetic resources in order to hedge risk in their agricultural decisions (so that they will retain a diversity of plant genetic resources), or when determining the initial set of plant genetic resources available to the "traditional farmer" and the forms of exchange (e.g., between traditional farmers) that might be available between harvests, and when considering the technology utilized by the traditional farmers which must be flexible enough to allow natural selection to play an important role in farmer's choice of crop varieties.

Let us assume that it is possible to institute a program of *in situ* conservation, i.e., there is some subset of farmers whose choice of crop germplasm on their land is made in response to the shifting environment. The germplasm that results from this method of operation then incorporates a flow of information, i.e., the crop varieties in use by this set of farmers will then contain traits and characteristics that are effective under currently prevailing environmental conditions. These favored traits and characteristics represent a flow of information from nature to the farmers in the *in situ* conservation areas. Then the modern agricultural sector is able to utilize this information to inform its choices of crop varieties throughout agriculture.

The solution to Problem (7.1'), by virtue of *in situ* conservation, represents a well-known approach to the use of information in decision making. This formulation of the decision making process is generally known as a *closed-loop* or *feedback* rule. It is a rule where the values of the choice variables depend upon the ongoing performance of the system under control* (Holly and Hughes Hallett 1989). The solution to a problem stated within the closed-loop format is normally a function (rather than an explicit set of values). That is, the solution is a process of information acquisition and utilization rather than a specific set of choices taken by reference to the information available at one point in time. *In situ* conservation therefore accords with the idea of a closed-loop method of decision making: it contemplates basing the decision in each period on the best information available *in the period in which that decision is taken.*

---

* The special case of a stationary function is normally described as a *stationary Markov strategy* (Cornes et al. 1995) which takes as its arguments the currently observed results from recent choices. $\bar{\mu}_t = 0_t(\bar{y}_t; \Phi\Pi) = 0_i(\Omega_t)$

There is no doubt that there is information arriving in each period that is potentially valuable in decision making regarding the control of modern agriculture. The amount of information provided is necessarily limited, however, by the size of the set of genetic resources in continued interaction with the environment and by the capacity to observe changes in performance by the decision maker. The cost associated with this information generating process is equal to the opportunity costs of the lands dedicated to *in situ* conservation, since the cultivation of sub-optimally performing varieties under sub-optimal technologies will reduce the expected present values of these operations.

To illustrate the nature of closed-loop decision making, consider the following simple example. Under an *in situ* conservation program, there will be a set of farmers who will devote a fixed proportion of the available land (c) to the cultivation of a diverse set of varieties (y_d) of a single crop. The quantities c and y_d are exogenously determined by the system of incentives established under the *in situ* conservation system. Meanwhile, by focusing only on yield information, the lands in the modern agricultural sector will be invested in the currently best-performing crop. Assuming that there is a relatively low level of output on the lands invested in conservation, aggregate agricultural output in period t is therefore:

Agricultural Output (with *in situ* conservation costs):

$$Y_t = (1-c) \cdot \left[E\left(y_{e_t}\right)^{\alpha_e} \right] \cdot \vec{p}_t$$

The decision rule in each period reduces to assigning the soil resources $(1-c)$ to the asset e which maximizes Y_t. A closed-loop decision-making process does this in a manner that makes maximum use of the information that is expected to flow into the system. Here we will focus on the use of the information flowing from nature, as derived from the lands invested in conservation (c). Therefore, looking forward one period and holding technology constant and neglecting preferences, output in $t + dt$ will be:

Closed-Loop Decision Making:

$$Y_{t+dt} = (1-c) \cdot \max\left\{ \left(\hat{y}_e\right)^{\alpha_e} ; \left(\hat{y}_e + \Delta\right)^{\alpha_e} ; \left(\hat{y}_e - \delta\right)^{\alpha_e} ; \left(\hat{y}_f\right)^{\alpha_f} ; \left(\hat{y}_f + \Delta\right)^{\alpha_f} \right\} \quad (7.2)$$

where

$$\hat{y}_f = \max_t \left\{ y_d \right\}$$

Equation (7.2) states that output in the modern agricultural sector will be produced by using the best available option from **either** the previous input variety *e*, potentially changed by depreciation or adaptation **or** the best variety *f* available from the set of diverse resources in period *t*, or a variety from that set that has recently been adapted to existing environmental conditions. This means that modern agriculture is able to rely upon the genetic resources within that sector so long as they produce the best yields, but that there are other sectors available if that is not the case. More importantly, the alternative sectors are simultaneously producing the information on the important traits and characteristics for adaptation while the environment continues to change.

For example, the usual pattern of use regarding a particular plant variety indicates that pest resistance will erode and render that variety economically nonresistant within 4 or 5 years; this rate of environmentally induced depreciation is represented by the third term in Equation (7.2) above. On account of this predictable rate of depreciation (and the unlikelihood of economically significant adaptations in a monocultural system), the alternative varieties in use in the conservation system begin to become relatively more attractive; this is represented by the fourth term in Equation (7.2). The conservation system operates as a "bank" of previously existing but inferior varieties. However, the single most important function performed by the conservation system is the capture of a flow of adaptations within that system; this is represented in the final term in that equation. It states that the *in situ* system will observe and make use of any important adaptation signaled within that environment. All that is required is the land use decision providing for the dedication of some amount of land to the cultivation of a wide range of diverse varieties. Then the desirable traits and characteristics identified within the diverse *in situ* system may be cycled into the more uniform modern agricultural sector on a systematic basis.

Therefore, *in situ* conservation is an approach that maintains a set of farming systems for the information that such systems will generate for the decision-making process. Each period decisions must be made concerning the maintenance of agriculture, and each and every farm practicing traditional and diversity-based agriculture acts as a *receptor* of information on the shifting of the natural environment. The greater the number of receptors in existence, the greater the likelihood that the information on the solution to the problems inherent in the current shifts in the environment will be available. *In situ* conservation represents an approach dedicated to the capture of this incoming information.

Ex situ *conservation as an open-loop strategy*

Ex situ conservation may be conceptualized as a very different form of decision making regarding the problems arising in modern agriculture. It is based on the idea that the solution to future problems is probably to be found in the set of currently existing genetic resources. Rather than base decision making on the capture and use of a flow of future information, the *ex situ* approach bases decision making on the optimal use of an already existing stock of information (represented by the already existing closely related varieties). In short, the two approaches are distinct solution concepts to the same problem, and both are necessary components of a complete solution to agricultural problems.

We will conceptualize *ex situ* conservation as a process in which the decision maker selects the set of genetic resources to be used in the maintenance of modern agriculture at a single point in time (t_0). The decision maker does this by selecting the optimal set of assets from the available genetic pool at this time and storing them for future use as inputs into the agricultural production process. This decision-making process is distinct from the previous one because it is based on the optimal use of the set of information already existing rather than the optimal appropriation of a flow of incoming information. The decision-making rule in this case may be stated as:

Open-Loop Decision Making:

$$\bar{u}_t = g_{t,t_0}\left(\Omega_{t_0}\right) \tag{7.3}$$

This is the usual formulation of an *open-loop decision rule*: The decision maker commits herself to a specific decision-making process across time based on a calculation procedure $g(\cdot)$ applied to a given set of information Ω available at some particular point in time t_0 (Holly and Hughes Hallett 1989). In this context the given set of information consists of the stock of genetic resources available for banking at a particular point in time. The irreversibility of genetic erosion imposes the restriction of a non-increasing set of genetic resources in storage over time (Frankel et al. 1995).

This is the form of decision making that is used when the supply of genetic resources is restricted to the use of gene banks. From the set of already existing varieties, a set is selected for conservation within the gene bank. This information set is then "frozen" at the time of collection. The remaining unbanked stocks of genetic resources are increasingly lost through displacement by modern agriculture. The flows of future information are lost by reason of the displacement of traditional agricultural land uses by modern agriculture. In short, *ex situ* conservation represents a decision-making process concerning the optimal use of the already existing stocks of information inherent in landraces and other stocks of genetic resources, and nothing more.

Optimal ex situ *conservation (open-loop decision making)*

Although *ex situ* conservation is necessarily an incomplete approach to the problem of agricultural depreciation, it remains an important component of the overall solution. The optimization of *ex situ* conservation concerns the selection of the optimal number and type of accessions to be cataloged. This will simply require the decision maker to consider the cost of the marginal accession against its expected benefits. The criterion used for the assessment of the expected benefits is crucial to the exercise. If the assessment is a purely open-loop form of decision making (i.e., neglecting the role of future information that may arrive), then the criterion utilized will simply be the maximization of current expected values.

This optimization problem may be discussed within the context of Equation (7.1′) above. Recall that this is a description of the problem for a decision maker interested in maximizing the expected value of agricultural output over time. In the context of gene bank management this is translated into the decision on how to maximize the current expected value of yield over a (potentially infinite) time horizon minus the social costs of the gene bank collection:

Optimal *ex situ* Conservation Objective:

$$\max_{m} \int_{t_0}^{\infty} e^{-\rho t} \cdot E_0(Y_t) - x_t(m) dt \qquad (7.4)$$

where $x(m)$ is the cost of an *ex situ* collection of size m which is subtracted from the expected output Y of each period t based on expectation formed in the start of the program E_0. This optimization problem is a very special case, because it states that the objective is to make choices at the current time that will apply for the duration of the program and that these choices will be informed solely by the information available at this time; current expectations and beliefs regarding dynamic processes will determine all that is done in the future. In this case, the entire exercise would be dominated by those choices that currently generate the highest production values or are expected to generate the highest production values.

The "open-loop" conservation problem is simply to choose the subset of the existing set of related varieties on which to expend financial resources for their conservation. As one possible choice, suppose that at time t_0 we cardinally rank all of the n varieties from the one with the maximum expected yield $\hat{y}_0$ to the one with the minimum yield $\hat{y}_n$ as in the following relation.

Ranking by "Expected" Yield Performance:

$$\hat{y}_h = (1 - f(h)) \cdot \hat{y}_0$$

where $f(h)$ is the density of the cumulative distribution function of yields over the crop spectrum at point h observed at the point of decision making.

What factors into the expectations concerning potential yields in this context, i.e., what factors are important in determining the relative ranking of genetic resources? Under open-loop decision making, it is only current performance or current expected performance that matters in this ranking. For example, given existing technologies and preferences but complete uncertainty regarding the future environment, those genetic resources that are closely related to the existing high-performers would probably be expected to be those most readily incorporated within existing agriculture. Therefore, a high ranking would be expected to be associated with good current performances or high relatedness to a good performer.

Once a ranking is determined, it is optimal for the decision maker to stop collecting where the present value of the yield of the last collected variety equals the marginal cost of its inclusion in the germplasm collection. In terms of the problem set forth in Equation (7.4), the optimal solution is that ordered subset of all crops n described by an optimal "stopping rule" (Simpson et al. 1996). This rule specifies at which point a search process should be terminated when information is costly.

Optimal *ex situ* Conservation:

$$B = \{m \mid V(m) - V(m - 1) \geq x'(m)\} \subseteq \{n\} \qquad (7.5)$$

where $V(\cdot)$ denotes the present value of the expected yield of agriculture given a stock of genetic resources of size m. This rule states a very simple idea, which is that it is optimal to stop banking genetic resources when the marginal value of the next accession is less than the marginal cost of its storage.

The crux of *ex situ* conservation is the concept of expectation. Since the exact values for the performance of any crop variety at any stage later than t_o is not known (due to the future information flows which are unknown in this decision-making process), the decision maker must instead use current expectations concerning future yields as proxies in the selection of crop varieties. Expected yields are influenced by movements in the decision-making environment which are predictable, but not those which are not. Hence, in respect to the shifting decision making environment described in the second section of this chapter, the *ex situ* decision incorporates only expectations in regard to the processes which move in systematic fashion, e.g., the anticipated depreciation of modern varieties and the anticipated shifts in societal preferences. The other systems which move in a less predictable fashion are likely to be ignored in this decision making process because complete uncertainties (e.g., equal probabilities of direction of movement) do not affect expectations. Since this decision is irreversible, any information arriving thereafter is irrelevant to the optimization program (Conrad and Clark 1987).

Optimal genetic resource conservation:
combined in situ *and* ex situ *strategies*

Optimal genetic resource conservation for food security in agriculture is a general problem composed of two parts: the first part concerns the optimal use of existing stocks of information (primarily for immediate yield improvements) and the second part concerns the optimal appropriation and use of future flows of information (primarily for the maintenance of current yield levels). For the dynamic aspects of the problems of agriculture, it is best to use a dynamic approach to decision making; this implies the use of *in situ* conservation for addressing the optimal appropriation of flows of information while *ex situ* conservation is used to optimize the use of existing stocks of information. In this section we describe the optimal combination of *in situ* and *ex situ* strategies for these purposes.

Optimal ex situ *conservation revisited*
In the previous section we described the optimal approach to *ex situ* conservation where the objective was the optimal use of currently available information in maximizing future yields. Of course, the primary application of this approach is in the decision making concerning the conservation and use of the existing stock of genetic resources. The rules developed there describe how those genetic resources that are expected to be most useful will be those that will be conserved. Given the correlation between relatedness and expected usefulness, it would be expected that gene banks would conserve large stocks of closely related genetic resources. This is in fact the case.

The only issue outstanding under this approach would be whether the expected benefits from any given variety would outweigh its cost of conservation; this would appear to be the approach used by other analysts (Evenson 1996). The issue to be considered now is whether this optimal *ex situ* approach addresses both sides of the conservation problem: optimal use of both current stocks and future flows of information. Does the conservation of large numbers of related varieties address both sides of the agricultural security problem?

The incremental value of in situ *conservation.*   The existence of future flows of information is equivalent to the existence of relevant uncertainty. Where uncertainty matters, closed-loop or feedback rules are always superior to open-loop policies since the former focuses on the use of any new information arriving over the course of the relevant time horizon (Karp and Newbery 1989; Holly and Hughes Hallett 1989); they provide an additional value attributable to future flows of information. This *expected value of information* is defined as the difference between the expected values of the maximized objective both before and after observing the information (Dasgupta and Heal 1979). This value coincides with the concept known as "quasi-option value" which is used to describe the value of delaying irreversible decisions if relevant information is expected to arrive between the present and the postponed point of decision making (Conrad 1980).

The expected value of information (EVI) in the case of crop genetic resources is expressed as the present value time integral of the difference between the closed-loop rule adopted through the use of *in situ* conservation methods (*c*) and the open-loop rule associated with *ex situ* conservation (*m*):

The Incremental Value of *in situ* Conservation:

$$EVI = \int_t e^{-rt} \left[(1-c) \max\left(E(\bar{y}_t(c))^{\tilde{a}} \right) - E_0 \left(\max\left(E(\bar{y}_t)^{\tilde{a}} - x(m) \right) \right) \right] dt \quad (7.6)$$

where E_0 denotes expectations formed in t_0. Equation (7.6) states that the incremental value of *in situ* conservation accrues by reason of the postponement of important decision making until that point in time when the relevant information has arrived. *Ex situ* conservation equates with the decision making at the time of storage rather than at the time when the nature of the relevant problems in agriculture is known. *In situ* conservation, on the other hand, awaits the arrival of the relevant information before the decision on genetic resource conservation is made. *In situ* conservation maintains a broader range of resources within an active environment, and then selects for use (and continuation into the next period) those genetic resources that nature itself reveals as the most significant. This allows for a step-by-step approach to decision making based on the actual information available at the time of the decision, as opposed to a path based on the information that is expected to arrive over the relevant time horizon.

What factors affect the value of *in situ* conservation? Three implications of Equation (7.6) are:

1. The expected value of information is an increasing function of the relevant time horizon since expectations formed at the outset of the program become increasingly less accurate estimates of the actual values at increasingly more distant points in time *t*; more information arrives with more time.
2. The expected value of information is clearly a function of the extent to which it is predictable that there is a flow of information but that the nature of that information is unpredictable; this was demonstrated in the second section.
3. The size of the *in situ* conservation determines both the current costs of and the future benefits from information flow, and therefore has ambiguous effects. In fact, determining the optimal magnitude is where the actual trade-off in *in situ* conservation lies.

Both *in situ* and *ex situ* methods are substitutes as mechanisms for conserving stocks of information but only *in situ* conservation appropriates and makes use of the future flow of information. Existing landraces may be preserved via either method although imperfectly, since *ex situ* conservation suffers as a stock conservation mechanism from losses due to storage problems and *in situ* conservation suffers from genetic drift. But only *in situ* conservation will aid in the identification of those traits and characteristics which are most important in the context of future environmental shifts. Although this is widely accepted as self-evident in the case of nondomesticated forms of biodiversity (Frankel and Soulé 1981; Frankel et al. 1995), the discussion about crop genetic resource preservation has not given this aspect sufficient attention.

In this section we have demonstrated that optimal genetic resource conservation must be conducted through the combined use of two distinct instruments: *ex situ* methods and *in situ* methods. This is because the problem of sustaining agriculture over any reasonable time horizon will require the use of both existing stocks of information and future flows of information. Agriculture represents the confluence of several dynamic systems, natural and social, and this dynamism must be incorporated within the decision-making process concerning the overall system. Otherwise the use of a entirely static approach to a clearly dynamic problem will be necessarily inferior. Both forms of conservation are clearly necessary in order to approach an efficient method of genetic resource conservation.

Optimal in situ *conservation*

The optimal amount of *in situ* conservation may be defined in an abstract fashion. Note from Equation (7.6) above that the expected value of information (EVI) is a function of the quantity of lands invested in *in situ* conservation (c). As more lands are placed into traditional farming regimes, there exist more "receptors" in the environment to detect and inform on the nature of environmental shifts. However, there is also implicit within Equation (7.6) an additional costliness to each additional conservation area; this is the opportunity cost of the forgone production from the marginal conservation area. Differentiating (7.6) with regard to the quantity c would yield the expression providing the decision rule for optimal *in situ* conservation; it may be stated as follows:

Optimal *in situ* Conservation:

$$EVI'(c) = y_c \qquad (7.7)$$

This decision rule merely states that the marginal piece of land placed into *in situ* conservation will be that piece which yields informational value just equal to the opportunity costs implicit in its removal from production. That is, the value of the land as a mechanism for generating information to sustain modern agriculture must be equal to the loss of production value involved in its removal from modern agriculture. The rule is based on one period's loss in production value because there is nothing irreversible about maintaining lands in *in situ* conservation; any lands retained in traditional varieties may be converted to modern varieties in the next period at the cost of one lost harvest.

There will be many factors that determine the optimal policy for *in situ* conservation. This is because there are many different factors that will influence the expected informational value of a given conservation area. This will depend, for example, upon its relative environmental uniqueness, its proximity to crop growing areas, and its relation to previously designated *in situ* conservation areas. Redundancy in any of these dimensions (environmental, regional, or agricultural) will reduce the informational value of the proposed reserve.

To take up our conception of *in situ* conservation in a manner analogous to the erection of a network of environmental information "receivers" (something like weather stations), it is apparent that the wider the range of genetic resources maintained at each site and the greater the number of sites, the greater the amount of useful information that will be received from this network. A rational system of genetic resource conservation will include an optimal system for the monitoring of such information, and hence it will include an investment in the maintenance of an optimal *in situ* conservation system.

Conclusion

This chapter has demonstrated the optimal set of policies required for the supply of genetic resources to solve problems of instability in modern agriculture. Such problems arise in a predictable fashion in agriculture because of its existence within a dynamic setting: natural, social, and technological systems continue to shift and introduce nondeterministic change within this context. In particular, agriculture represents the management of the biological world for the supply of necessary goods and services to human societies, yet the motion within the biological world attributable to evolution is both continual and nondeterministic.

The particular problem that such dynamism introduces concerns the continual erosion of the resistance of the varieties used within the modern agricultural system. The solution for this problem concerns the ascertainment and implementation of new forms of resistance to the predictably recurring problems of pest invasions. An industry exists that undertakes this task, and the sole issue that concerns government is to ensure that an optimal supply of germplasm is maintained in order to supply this industry into the future.

This chapter has argued that this problem should be conceived as a problem of optimal decision making concerning information acquisition under conditions of uncertainty. The maintenance of the existing varieties used in agriculture is equivalent to the conservation of stocks of information accumulated within those varieties by reasons of centuries of evolutionary and human selection. This is an important task, and it may be accomplished (imperfectly in either case) via either *ex situ* or *in situ* methods of conservation. The maintenance of lands in traditional agriculture (i.e., a range of diverse varieties subject to natural and farmer selection) represents a mechanism capable of generating a flow of future information concerning existing conditions in future environments. This essential task may be accomplished solely through the mechanism of *in situ* conservation.

This chapter has therefore demonstrated that the optimal approach to the solution of the problems of food security in modern agriculture will encompass the use of both methods of conservation: *in situ* and *ex situ*. It has further demonstrated the considerations which should be determinative of the extent to which each method is undertaken. Most importantly, it has demonstrated the underlying nature of genetic resource conservation (the management of information) and the fact that it has two distinct components: the regulation of past stocks of information and the regulation of future flows.

The conservation of plant germplasm for agriculture is a classic instance of this particular policy problem. Modern agriculture is now a massive construct sustained by a small but crucial flow of information. This illustrates the hubris of human society's static conception of previous agricultural successes; any gains (in terms of increased yields) must be perpetually defended. The existence of a large portion of the biosphere invested in a small number of species (namely, humans and their associated domesticated/cultivated species) *does not* indicate an inherently stable system. This situation in fact represents an opportunity for exploitation by other biological organisms: successful invasion implies massive gains in fitness. Evolution will constantly and perpetually introduce new variants of pests and parasites for the invasion of this human domain.

Humans must likewise constantly and perpetually defend their domain against these forces. This is an important part of the task that society has set for the plant breeder/agricultural industry. Agricultural R&D should be seen as a dynamic contest between human societies and nature. Industrial R&D processes represent human society's attempts to innovate winning strategies at a rate faster than the biosphere is able to evolve ones to defeat them. Pest resistance is a problem that never goes away. Although the usual cycle for product development in the plant breeding industry is about 10 or 11 years, the developed resistance characteristic is often viable for only 4 or 5 years (WCMC 1996). Hence a continual cycle of breeding for new resistance characteristics is required.

The policy maker's objective is to ensure that there will continue to be a flow of information capable of providing solution concepts for the predictable problems that will arise in modern agriculture. Private industries will work to resolve the immediate problems with the resources available, but they do not have the long-term perspective necessary to provide for the maintenance of agriculture into the distant future. The objective of policy makers in this context must be to provide the mechanism which will sustain this flow of information. The object is not necessarily to conserve a particular set of germplasm, but rather to conserve a system which will provide the correct set of germplasm for the solution of future problems. All of these factors point to the necessity of the creation of an optimal mechanism for *in situ* conservation.

References

Arrow, K.J. 1962. Economic welfare and the allocation of resources for inventions. In *The Rate and Direction of Inventive Activity,* R.R. Nelson (ed.). Princeton, NJ: Princeton University Press.

Brush, S.B., E. Taylor, and M. Bellon. 1992. Technology adoption and biological diversity in Andean potato agriculture, *Journal of Development Economics* 39:365–387.

Conrad, J. 1980. Quasi-option value and the expected value of information, *Quarterly Journal of Economics* 94:813–820.

Conrad, J. and C. Clark. 1987. *Natural Resource Economics. Notes and Problems.* Cambridge: Cambridge University Press.

Cornes, R., N. Van Long, and K. Shimomura. 1995. *Drugs and Pests: Negative Intertemporal Productivity Externalities.* McGill University. Mimeo.

Dasgupta, P. and G. Heal. 1979. *Economic Theory and Exhaustible Resources.* Cambridge: Cambridge University Press.

Ehrlich, P., A. Ehrlich, and G. Daily. 1993. Food security, population and environment, *Population and Development Review* 19:1–32.

Evans, L.T. 1993. *Crop Evolution, Adaption and Yield.* Cambridge: Cambridge University Press.

Evenson, R.E. 1996. Valuing genetic resources for plant breeding: hedonic trait value, and breeding function methods. Paper presented at FAO Symposium on Valuing Genetic Resources, 13–14 May 1996. Universita 'degli Studi di Roma "Tor Vergata," Rome.

Evenson, R.E. and D. Gollin. (1997). Genetic resources, international organizations, and improvement in rice varieties, *Economic Development and Cultural Change* 34:471–500.

Fabella, R. 1989. Separability and risk in the static household production model, *Southern Economic Journal* 55:954–961.

Fafchamps, M. 1992. Cash crop production, food price volatility, and rural market integration in the third world, *American Journal of Agricultural Economics* 74:90–99.

FAO. 1996. *Draft Report on the State of the World's Plant Genetic Resources* (Full Background Documentation). Mimeo.

Frankel, O., A.H.D. Brown, and J. Burdon. 1995. *The Conservation of Plant Biodiversity.* Cambridge: Cambridge University Press.

Frankel, O. and M. Soulé. 1981. *Conservation and Evolution.* Cambridge: Cambridge University Press.

Goeschl, T. and T. Swanson. 1996. Market imperfections and crop genetic resources. Paper prepared for the International Plant Genetic Resource Institute. Rome: IPGRI.

Harlan, J. 1975. *Crops and Man.* Madison, WI: American Society of Agronomy.

Hazell, P. and R. Norton. 1986. *Mathematical Programming for Economic Analysis in Agriculture.* New York: Macmillan.

Holly, S. and A. Hughes Hallett. 1989. *Optimal Control, Expectations and Uncertainty.* Cambridge: Cambridge University Press.

Karp, L. and D. Newbery. 1989. Intertemporal consistency issues in depletable re-sources. CEPR Discussion Paper No. 346.

National Academy of Sciences. 1972. *Genetic Vulnerability of Major Crops*. Washington, D.C.: National Academy of Sciences.

Orians, G., G. Brown, W. Kunin, and J. Swierzbinski (eds.). 1990. *The Preservation and Valuation of Biological Resources*. Seattle: University of Washington Press.

Plucknett, D. and N. Smith. 1986. Sustaining agricultural yields, *Bioscience* 36(1):40–45.

Plucknett, D., N. Smith, J. Williams, and N. Anishetty. 1987. *Gene Banks and the World's Food*. Princeton NJ: Princeton University Press.

Simpson, R., R. Sedjo, and J. Reid. 1996. Valuing biodiversity for use in pharmaceu-tical research, *Journal of Political Economy* 104(1):163–185.

Swanson, T. (ed.). 1995. *The Economics and Ecology of Biodiversity Decline*. Cambridge: Cambridge University Press.

Swanson, T. 1996a. Global values of biological diversity: the public interest in the conservation of plant genetic resources for agriculture, *Plant Genetic Resources Newsletter*, No. 105.

Swanson, T. 1996b. The impact of intellectual property right systems on the conser-vation of biological diversity. Paper prepared for the Secretariat of the Conven-tion for Biological Diversity.

Swanson, T., R. Cervigni, and D. Pearce. 1994. *The Appropriation of the Benefits of Plant Genetic Resources for Agriculture*. Commission on Plant Genetic Resources. Rome: FAO.

World Conservation Monitoring Centre and Faculty of Economics, Cambridge Uni-versity. 1996. *Industrial Reliance Upon Biodiversity*.

chapter eight

The Cultures of the Seed in the Peruvian Andes

Tirso A. Gonzales

Introduction

In contemporary societies the seed is embedded, in general, within two types of cultures: Western cultures and indigenous peoples' cultures. Within the former culture we find what has evolved as the Culture of the Commercial Seed. Indigenous peoples' cultures, so-called "traditional," "primitive," "reluctant to change," or "tribal" cultures, have developed what I denominate a sophisticated culture, the Culture of the Native Seed. This chapter explores the cultures of the seed as a framework to understand that the seed does not have the same meaning, or play the same role, in Western contemporary dominant agriculture as in indigenous peoples' agriculture.

This recognition should contribute to the understanding that the term *in situ* is not a universal one; thus there is not just one strategy of *in situ* conservation. In the indigenous world there are many conservation strategies. These are embedded within the diversity of indigenous peoples' agriculture. In the Americas (North, Meso-, and South) the current indigenous population is around 42 million people, with a total of 900 languages. Measured by language as well as by biological and cultural diversity, it is indigenous peoples who show a high correlation between cultural and biological diversity. From a global view, Mexico, Brazil, Peru, Colombia, and Ecuador are among the 12 countries with the highest biological diversity (Durning 1992). In Peru, a country noted for having one of the highest levels of agrobiological diversity worldwide, 51 ethnic groups — a population of more than 9 million people (around 40% of the total population in 1992) — practice a diversity of *in situ* conservation strategies. The direct historical and contemporary contributors of such great biological diversity are the indigenous peoples of Latin America. In the Andes, it is the Quechua and Aymara

Originating Peoples who are the primary nurturers of agrobiodiversity. This chapter describes the efforts of a unique indigenous non-governmental organization, the Asociación Bartolomé Aripaylla (ABA) based in the indigenous community of Quispillacta, in the southern Andean state of Ayacucho, Peru. In contrast to governmental and other non-governmental organizations, ABA fosters on-farm conservation of agrobiodiversity within the framework of cultural affirmation (Grillo 1993). Cultural affirmation has as its cornerstone the *chacra* (something more than just the "peasant's plot"). As I will present later, the strengthening of the *chacra* regenerates life (agrobiodiversity included) and the community as a whole.

The chapter concludes by suggesting that the Andean Culture of the Native Seed, embedded in the Andean indigenous communities, offers unique possibilities for promoting *in situ* (on-farm) conservation of crop genetic resources in the Andes. Adoption of such practices implies learning from Andean originating peoples in their own terms (Ballón et al. 1992). Furthermore, it calls for the resolution of the Indigenous Question, that is, the struggle of Andean and Amazonian indigenous peoples for self-determination, control over their territories and resources — natural, intellectual, and communal. Understanding the cultures of the seed also presents the indissoluble link between biological and cultural diversity, making our ecological concerns the natural province of cultural and political realms.

The place of in situ *conservation*

The theory and implementation of *in situ* conservation of plant genetic resources is relatively recent. The strategy has been determined by contemporary Western societies and institutions that seek to conserve and manage plant diversity in a scientific fashion. As Browning reminds us, "In the early 1970s, few scientific or government leaders accepted the idea of *in situ*, or on-site, conservation of crop genetic resources. But some capable scientists did so when the idea was presented to them logically" (Browning 1991:59). This concept responds to a logical question of scarcity of local crop genetic diversity and is closely associated with contemporary Western scientific approaches to nature in highly industrialized countries in the North (Europe, North America, Japan, and Taiwan included). These countries are well endowed with scientific infrastructure and are financially rich but poor in plant genetic resources (Kloppenburg 1988).

Throughout the past century and most of this century, within key dominant U.S. institutions and agencies, the central connection between agricultural development and its impact upon Native American agricultures has been missing (Haynes 1985). Twelve years ago Haynes noted that "in the U.S., industrial agriculture has tended to be monocultural both in the sense that it is intolerant of alternative cultures and in the sense that it relies heavily on monocultural cropping systems" (Haynes 1985:1) Monocropping is associated with the international agricultural phenomenon known as the "Green Revolution," which in practice is nothing else than the export, since the

1960s, of U.S. modern science-based and corporate-biased agriculture. The contemporary "Green Revolution" has favored the narrowing of the genetic base in the northern industrialized countries (Griffin 1974; Pearse 1980). Equally important is what Nabhan noted in 1985: "today, less than 20,000 Indian families in the U.S. continue farming, and probably only a small percentage of these grow the heirloom crops of their forefathers" (Nabhan 1985:15).

"Green Revolution" techniques have been widely applied in "Third World" agriculture since the 1950s via the transfer of technology model of agricultural research. Pimbert notes this model, typical of both national and international agricultural research systems, has been instrumental in expanding the application of "Green Revolution" agriculture in the "Third World."

> Reductionist research, high input packages and top down extension have led to successes: in the uniform and controlled conditions of industrial and green revolution agriculture, they have raised output per unit of land. The simplifying tendencies of reductionist science have meshed well with the ecological and social simplicity of standardized, specialized farming systems (Pimbert 1994:20).

The Consultative Group on International Agricultural Research (CGIAR), the National Agricultural Research and Extension System (NARS), and agricultural universities form the cluster of institutions associated with the "Green Revolution" technologies. By different means and degrees, "Green Revolution" technologies created in the U.S. through the land grant college system, non-land grant universities, and agribusiness have been adopted by major international development and finance institutions and "Third World" states to develop and modernize agricultural production in the "Third World." Included among the sectors to be modernized are the "backward" indigenous peasant farmers and their "primitive" agricultural systems.

The Western theory and practice of *in situ* conservation of plant genetic resources is foreign to indigenous peoples' cultures and agricultural practices. *In situ* conservation might be a valid concept in the Western world (Maxted et al. 1997). It is not, however, universal. In the context of "biodiversity" and "conservation," Pimbert provides a relevant illustration.

> There are multiple perspectives on what constitutes "biodiversity" and "conservation." Biodiversity as a word and concept has its origins in the field of conservation biology. It is a Western category that refers to the variability among living organisms and the ecological complexes of which they are part; this includes diversity within species, between species and of

> ecosystems. But for indigenous peoples, "biodiversi-
> ty" is often best understood as, for example, "Mother
> Nature." Also the Western concept of "conservation"
> generally cannot be translated — indigenous equiva-
> lents are often framed in terms of "respecting nature"
> or "taking care of things." Where Western trained pro-
> fessionals might speak of conservation as a separate,
> discrete and special activity, indigenous peoples often
> view this fragmentation of the world as strange. "Tak-
> ing care of things" is not disconnected from the rest of
> their activities — it is part of indigenous peoples' way
> of life and livelihoods (Pimbert 1994:3).

In situ conservation is part of specific recent contemporary policies and
disciplines such as agro-ecology, conservation biology, and botany. Despite
the good intentions of its practitioners, those Western disciplines and their
respective theories share and are grounded in similar theories of the self or
ontology, theories of knowledge or *epistemology*, and theories of the universe
or *cosmology*. Lack of understanding of such fundamental differences may
lead to undermining indigenous agricultures and agrobiodiversity, as well
as misguide the allocation of scarce financial resources. More importantly, it
sets the foundation for continued misunderstanding of how and under what
conditions agrobiodiversity is nurtured in the indigenous world. *In situ*
conservation of crop genetic resources is part and parcel of a particular type
of Culture of the Seed, that of the Commercial Seed. By contrast, in the
indigenous world, life is produced, reproduced, and enriched within the
Culture of the Native Seed, which suggests that the respective cultures of
the seed are embedded within substantially different cognitive frameworks,
environments, rituals, histories and stories, peoples and practices.

In situ *conservation and the Indigenous Question: toward an integral approach*

It is necessary to question the way in which major international development and funding institutions — located mainly in the industrialized North
— study biodiversity. Research is not the final outcome of Western investigation (Buttel 1993). Western institutions propose alternatives regarding the
South's biodiversity — in particular crop genetic resources — without *full*
consideration or consultation with indigenous peoples (indigenous peasants, tribal people) and local communities, who have historically and continue today under oppressive regimes to know, manage, produce, and reproduce the genetic "raw material" that will be used by the geneticists and
exploited by the seed industry from the North. Further, as with contemporary modernization theories, the dominant environmentalist approaches to
crop genetic resources do not seriously address the Indigenous Question.

The dominant conservative environmentalist approaches limit our under-standing of the Indigenous Question and its connection with Nature and Culture. The Indigenous Question brings to the forefront what major inter-national development and funding institutions, states, non-governmental organizations, political parties, and government organizations have, within the context and implementation of capitalist modernization, tended to seri-ously neglect. The defense of the collective rights to biodiversity (agrobiodi-versity included) cannot be separated from the collective rights of indige-nous peoples (Stavenhagen 1990).

Both *in situ* and *ex situ* conservation are contemporary strategies which the Western world proposes to counteract genetic erosion. With the failure of *ex situ* strategies to capture evolutionary processes, development approaches are now considering the possibilities of *in situ* conservation (Oldfield and Alcorn 1991). Current *in situ* conservation strategies (e.g., farmer curator system, Protected Areas) are embedded in Western geopolit-ical and economic interests. Such strategies are too narrow, abstract, or naive when related to indigenous peoples. Dealing with agrobiodiversity and indigenous peoples requires an integral view (Grillo 1993; Richards 1993). Such an approach implies the consideration of the indigenous cultural and environmental context in which agrobiodiversity is produced, reproduced, and enriched.

The Cultures of the Seed: a working definition

For any contemporary strategy of *in situ* conservation, it is crucial to recog-nize the Cultures of the Seed and how they have unfolded within contem-porary capitalism. There are two types of seed: the commercial seed and the native seed, the former embedded within contemporary Western culture and capitalist agriculture, and the latter embedded within the rich diversity of indigenous agriculture. The native seed is the precondition of the commercial seed. The commercial seed — mostly under the control of transnational organizations — is the outcome of scientific manipulation in laboratories and test plots; it narrowly privileges some genetic traits. The native seed is the outcome of a nurturing process, embedded within non-Western cosmol-ogies, between indigenous peoples, the seed, and all other living beings. The concept of Cultures of the Seed is an analytical tool that highlights the fact that the seed is neither a simple commodity that plant breeders can manip-ulate in a lab nor something we buy at a seed store; nor does the seed evolve in a cultural or biological vacuum. The Cultures of the Seed implies specific cosmological views and cognitive models, diverse technological strategies and ecosystems as well as substantially different types of social, religious, and productive organizations (Chambi and Chambi 1995; Machaca 1993; Kloppenburg 1988; Descola 1989).

We see the seed's multiple dimensions most vitally in the intersection between society and nature, between cultural and biological diversity. Through the heuristic construct of the Cultures of the Seed, I make explicit key relationships between the seed and major contemporary global issues such as development. Historically, the term Cultures of the Seed refers to the intersection between capitalist society and nature (for the commercial seed) and between indigenous peoples, local communities, and nature (for the native seed). How does each culture of the seed relate to or appropriate nature? How is the seed cultivated in those two cultural settings? These relationships and the question of appropriation point to how crucial it may be to recognize the distinct characteristics of both cultures for the future of agriculture as well as the design of *in situ* conservation strategies.

Development and the Cultures of the Seed

In this century, and especially since the 1940s, the Latin American state has implemented specific policies in pursuit of modernization and development in the countryside. Latin American indigenous peoples, and their respective agricultures, have been the *object* of modernization. Within modernization blueprints, the expectation was (and still is) that indigenous peasants in contemporary Latin America would become modern, business-minded farmers, moving forward to take advantage of the supposed benefits provided by the applications of contemporary Western science to agriculture.

The production of the commercial seed requires the "genetic material" encoded in the native seed — both "(semi) wild" and "(semi) domesticated" — located mainly in the indigenous peoples' plots of the South. Native seed is rooted mainly within the diversity of indigenous cultures. Moreover, it grows in a mutually nurturing relationship within a particular cultural and ecological setting. For cultural, political, environmental, and social reasons, what might a comparison between those two types of cultures of the seed highlight for future research?

The culture of the commercial seed has been supported and strengthened by means of major international rural development projects and the agricultural research, education, and extension system (Gonzalez 1986; Jennings 1988; Oasa and Jennings 1982) with strong linkages to agribusiness corporations. This system is part of the cluster of institutions that reproduces and generates the Culture of the Commercial Seed at the international, regional, national, and local levels. Throughout this century, governmental and nongovernmental institutions, academics, and intellectuals involved in "modernization and development" have not been able to understand the complexity of indigenous peoples (Adams 1994). Furthermore, they have been unable to propose an alternative resolution to the Indigenous Question in the "Third World."

The Western debate on biodiversity in general, and plant genetic resources in particular (*in situ* conservation included), provides a unique opportunity to illustrate the ways in which nature, production, and culture not only intersect, but are inextricably interwoven. This intersection is not necessarily the same within Western and non-Western cultures (Latour 1993) and their respective agricultures. While in the Western world, concrete historical, cultural, and political processes have led to the analytical separation of these three spheres (Latour 1993), in the Andean indigenous world there seems to be no such separation (Gonzales et al. in press; Grillo 1993).

Nature, production, and culture: The Culture of the Commercial Seed

Since the mid-1950s modern agriculture (monocrop agriculture) aggressively invaded the Latin American fields and the peoples working in agriculture. The penetration of modern agriculture (capital, roads, synthetic chemicals, machinery) in Latin America has not been homogeneous. This is illustrated not only by the differentiated mechanization of agriculture and the adoption of agricultural inputs (knowledge, synthetic chemicals, capital), but also by the presence of different types of farmers and farming practices.

Today this type of agriculture is recognized by many as not only unsustainable, but also as a major contributor to the global environmental crisis (Redclift 1987; NRC 1989). There is a rapid growth of discourses pointing to the global proportions of the crisis, which stress the loss of biological diversity (Colchester 1994). These discourses are mainly framed within industrialized countries and the major international development and funding institutions. Within the current trend of modernization and development, these countries and institutions give evidence of their biased policies toward the protection or "preservation" of biodiversity at a global level (WRI, IUCN, UNEP 1992). This tendency responds to the interests of nation-states and corporations, to the neglect of indigenous peoples who have historically maintained cultural and biological diversity.

Nature, production, and culture: The Culture of the Native Seed

The culture of the native seed is fundamentally different than the culture of the commercial seed. Within the former culture, the seed's journey unfolds in fundamentally different ontological, epistemological, and cosmological scenarios to those of the contemporary Western world — "different digestive processes" (Grillo 1993). One culture, by principle, is oriented to nurture and enrich every expression of life. The other moves away, ever more, from the principle of nurturing. Within the culture of indigenous peoples, we find viable alternatives to the current ethnic and environmental crises. These alternatives are proposed in the work of academics and indigenous intellectuals, as well as in organizations directed by and for indigenous peoples. Cumulatively, this work points to a paradigm of development substantially different from the current and dominant one. Alcorn raises a challenging thesis that not only questions the current "Northern" biodiversity debate but puts indigenous rural communities (generically called peasant communities) at the center of such a debate:

> I believe that careful study would confirm my field-
> based observation that the majority of the Earth's

> biodiversity is not held in patrolled wildland reserves,
> but in landscapes managed by local people. Most of
> the world's biodiversity is on lands and waters used
> by traditional peoples marginal to the global economy
> (Alcorn 1991:318).

In sum, much of the current and rapidly growing international debate on genetic resources and biodiversity, in particular the one generated in the "North," tends to "environmentalize" issues. That is, current discourse privileges strategies within this "environmentalist" framework, rather than focusing on the population outside or within targeted regions, which co-evolve with these environments. In the North as well as in its state and agro-industrial counterparts in the South, the issue of crop genetic resources is compartmentalized and analyzed in a technocratic and scientific fashion. The major emphasis is not on the relation of nature and culture, but on patches of nature and short-term economic gains. Such scientific segmentation and abstraction tends to perpetuate a dangerous disregard for the people who manage, reproduce, nurture, and experiment with native crops in rural areas.

This situation requires further analysis in order to propose more suitable and enduring ways of coping with and reproducing biological and cultural diversity. Until international development and funding institutions as well as the state recognize, in theory and practice, the key connections between culture (cultural diversity), production (indigenous or capitalist agriculture), and nature (agrobiodiversity), most of the influential propositions of the development apparatus with respect to biodiversity will have a short-term effect. Both biodiversity and indigenous peoples will continue to be over-looked by the dominant institutions and peoples who carry out Western modernization processes in regions such as Latin America.

The Cultures of the Seed in the Peruvian Andes

Over a period of 8,000 years, a patient process of Andean indigenous agricultural experimentation opened the way to the domestication of plants and animals. As a result, a rich diversity of native plants and animals has been adapted to the unique ecological conditions of the Andes. The Andes holds 84 out of the 103 ecosystems identified in the world, and it is considered one of the eight principal centers of crop genetic diversity in the world (ISNAR 1987).

Contemporary sociopolitical and ecological background

In 1978, the indigenous population of Peru numbered over 6 million, 4 million of whom live in rural *Communidades Campesinas*, or peasant communities, which form the basis of social organization in the Peruvian Andes and are officially recognized by the national government. These communities manage approximately 29% (8.6 million hectares) of the total agricultural and pasture land in the country and are considered the most important social, economic, and cultural institutions in the rural area (Matos Mar 1991).

Although indigenous communities throughout Peru vary greatly in terms of access to resources, geographic location, degree of internal organization, cultural traditions, and market relations, some generalities do exist. These communities, for example, represent the poorest segment of the population, garnering only 2 to 4% of the national income (Gonzales 1996). Most Andean indigenous communities are situated between 2,000 and 4,000 meters above sea level in the Quechua and Suni regions. Across this range of altitudes, Quechua and Aymara indigenous people experiment with, produce, reproduce, and nurture their native seeds in the Andean *chacra* — a peasant's field of 1 to 2 hectares. From these small plots scattered throughout the Andes, and at local and regional Indian markets, the "gene hunters" (Juma 1989) — explorers, plant collectors, and plant breeders from both public and private national and international agencies — collect native seed varieties. Without this wealth of genetic material of Andean roots, tubers, and grains, the Culture of the Commercial Seed would not be capable of reproducing itself and generating profits.

The Culture of the Seed or the Culture of the Phenotype in the Andes?

The seed, within the Culture of the Native Seed, interacts with specific local sociocultural and environmental situations, or in Andean terms the local *Pacha*. In the case of the Andes, Van der Ploeg (1993:218) suggests that the local practice of potato selection and amelioration "allows for a step-by-step improvement of different phenotypical conditions — steps that in their turn can follow, for instance, the demographic cycle within the farming family [peasant family] or the logic of patterns of co-operation within the communities." The Andean indigenous practice of potato selection and amelioration tends to be reorganized by the introduction of Western scientific "breakthroughs" of so-called modern "improved" varieties. This reorganization takes place at the level of the agricultural calendar and the social and religious practices embedded in it (Van der Ploeg 1993). Scientific plant breeding, notes Van der Ploeg,

> demands a sudden and complete repetition of specified requirements in farmers' plots [and] time is converted from a basically indiscrete into a discrete category, the labour process changes from the skill to confront and exploit specific circumstances, to the skill of applying general and standardized procedures to circumstances that are to be seen as more adverse the more they are specific (Van der Ploeg 1993: 219).

Thus, at the level of the community's Culture of the Native Seed, there is a major conflict between the requirements of local peasant practice of potato selection and amelioration, and the scientific plant breeding routinely practiced in the labs and experimental stations of national and international agricultural research centers, removed from the seed's natural and domesticated environment and daily peasant life. A similar situation is shared by foreign, national, and regional universities whose work during this century has been mainly at the genotype level.

Local Quechua and Aymara knowledge, labor, technology, and ritual mediate the relation between Andean people and Andean nature. For the Andean, labor is not solely for the production of use-value, nor is it considered a simple economic activity. Work is not an individual activity and the world is not an object to be manipulated for some extractive activity (Van Kessel and Condori 1992).

Different actors, interests, and alternatives for the Andes

Two major agencies central to Andean indigenous agriculture are the Peruvian state and the International Potato Center (CIP). During the 20th century, the concern of the state for crop genetic resources, reflected in specific policies regarding genetic resources, was limited and weak (Gonzales 1987, 1996). Although the government increasingly expresses concern about genetic resources, "wild " and "domesticated" crops, and medicinal plants, it does not have a coherent program for *in situ* conservation. Furthermore, this concern does not necessarily translate into improved regulations to protect the rights of indigenous peoples and local communities over their resources — natural, intellectual, or communal. On the contrary, Andean Pact trade regulations might undermine national laws by preempting them. Andean Pact members have already agreed that any established regional accord overrules any national law of the country members. Decision 291 of the Andean Pact refers specifically to patents for crop genetic resources (Acuerdo de Cartegena 1991). At this time Peru has withdrawn from the Andean Pact.

CIP aims to rescue Andean plant germplasm through the Andean Eco-regional Initiative, which will focus on two areas: "(1) biodiversity as a resource for future agricultural production and prevention of deterioration of the agricultural land base, [and] (2) agricultural policies that affect the maintenance and preservation of agricultural resources" (Zandstra 1993:30). One important component of the initiative is the Andean Root and Tuber Crops Project, funded by the Swiss Technical Cooperation Agency (COTESU) and coordinated by CIP. CIP will collect not only Andean potatoes and sweet potato germplasm, but also the wide spectrum of Andean roots and tubers — what I denominate "the last Andean frontier." The role of "small scale" Andean farmers and diverse ethnic groups within this project framework is still unclear. The specific research to be conducted will be determined by the Consortium for the Development of the Andean Eco-region (CONDESAN) a consortium of institutions from Peru, Bolivia, Ecuador, and Chile, with support from the national agricultural research system, and non-governmental organizations (local, national, and regional). Andean indigenous "peasants," the *object* of the initiative, have been excluded as collaborators (subjects in their own terms) in program planning and implementation.

The Andean cosmology and the chacra

Agrobiodiversity and *in situ* conservation are foreign terms to the Andean worldview. The local *pacha* (locality) is central in the Andean cosmology. It encompasses a greater diversity than the term biodiversity. In the Andean cosmology, nurturing, dialogue, and regeneration have special meaning (Greslou 1991; Grillo 1993). Everything is alive — a mountain, a rock, water, women, and men — and everything is incomplete. The Andes is inhabited by living beings that are incomplete. This allows for dialogue and reciprocity in all aspects of existence. All beings are persons in a relationship of equivalents and mutual nurturing (*crianza*). At the local level, the *chacra* is the place where the dialogue among beings occurs. It forms the basic cornerstone of Andean life (Figure 8.1). The practice of agri-culture represents the dialogue which occurs between diverse beings within the natural collectivity, the *ayllu*, the local *pacha*. At the center of the *ayllu* we find the *chacra(s)*, and each of them is unique: a major factor underlying the centrality of locality. The *ayllu* is a kinship group, but it is not reduced to human lineage, but rather gathers each member of the local *pacha* (local landscape) (Grillo 1993). The *ayllu* is found in the local *pacha* (local landscape) where the three components that comprise the natural collectivity live. The *pacha* is characterized by being animated, sacred, variable, harmonious, diverse, immanent, and consubstantial (Figure 8.1).

To initiate a project of *in situ* conservation in the Andes implies the improvement of the Andean environment, or the local *pacha*; the *sine qua non* condition is the strengthening of the Andean natural collectivity (Figure 8.1). The "ecological rationality of peasant production" (Toledo 1990) makes the Quechua and Aymara communities the most adequate, and the most irreplaceable, ecological units (Mayer 1994). The *chacra* may be construed in its relationship with agrobiodiversity, crop germplasm conservation, crop genetic erosion, and *in situ* conservation. In terms of plant germplasm diversification and conservation, the indigenous community's particular way of "doing *chacra*" reveals the uniqueness of Quechua and Aymara local agri-culture.

In the vast literature on agricultural development and in the ideology and praxis of the international and national apparatus related to the agricultural research system (NARS), "traditional" farmers are "invisible." As Richards (1993:61) notes, indigenous "local knowledge is often or mainly outmoded and something to be replaced." As Van der Ploeg suggests, Andean indigenous Andean farmers and the Andean highlands are made "invisible." "Invisibility seems to become especially reinforced when all your careful attention and love for the land are at once declared insignificant by

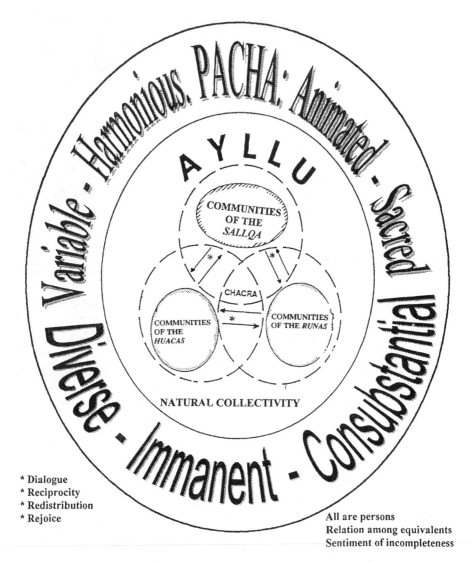

Figure 8.1   The Andean Ayllu (after Grillo 1993).

the introduction of general schemes to be followed in production and by the introduction of 'miracle seeds'" (Van der Ploeg 1993:221).

A lack of understanding of Andean culture (worldview) combined with a strong bias toward a science-based agriculture seem to be the factors working against the production, reproduction, and strengthening of Andean cultural and biological diversity. Furthermore, the "oneness of the modern world" (Richards 1993) is reflected in the literature on agricultural and rural development. This literature is not prone to acknowledging cosmovisions other than the contemporary Western cosmovision. The "invisibility" effect

(Van der Ploeg 1993) then leads to the disregard for indigenous peoples and their current and potential contribution to cultural and biological diversity (Lumbreras 1992).

Enriching seed diversity in Quispillacta

The community of Quispillacta is located in the district of Chuschi, province of Cangallo, in the department of Ayacucho. A recognition of Quispillacta, of the native Quechua language, and of the specific relation that the community members of Quispillacta hold with their land and their surrounding "living" environment constitutes a substantial difference from the cosmology in which modern agriculture is embedded. The territory of Quispillacta comprises 21,680 hectares (97% natural pastures, 2% arable land). The local geography is highly uneven; each ecological floor is different. Two river basins — the Cachi river basin and the Pampas river basin — hold a diversity of tubers and cereals.

Quispillacta is organized at three levels: *ayllu, barrio,* and the community. The extended family coalesces in the *ayllu,* which is strengthened through collective works and ritual festivities. Various *ayllus* make up a *barrio.* Quispillacta has a population of 6,000, grouped into 1,200 families (1996). The uniqueness and wide ecological, topographic, and climatic diversity and variation of the Andes demands a close relationship between the peasants and the *chacras* they manage at different altitudes and ecological niches. Weather prediction and standardization, as a component of science-based commercial agriculture, is not as easily applied in Andean agriculture. A finer and richer dialogue between the Quispillactans and their living environment facilitates their performance in their *chacras.* Learning to "read and understand" a number of signals (e.g., smell of the air, position of stars, bird and insect activities) is the key to nurturing life — crop genetic diversity included — in the Andes.

Quispillacta's people base their social and productive activities on the agricultural calendar. Their agricultural activities are intimately intertwined with specific religious festivities, rituals, and social relations. The life cycles of the human community, nature, and the deities interact in a process of dialogue and reciprocity among equivalents. This is illustrated below with the process of incorporating a seed to a family, *ayllu, barrio,* and community.

The incorporation of a seed to a *chacra* involves a number of steps (ABA 1993; Machaca 1993; Van der Ploeg 1993). The steps depend on the type of crop, the region, the agricultural calendar, the ecological niche, and the socioeconomic situation of the household units. The state of the Andean cosmovision, local knowledge, and the Quechua and Aymara's specific reading of the natural environment, such as the position of the stars and the nesting patterns of certain birds (Chambi and Chambi 1995), also play a role.

The incorporation of the seed into the family of crops in Quispillacta follows a number of rituals and trials over the course of several years

(Table 8.1). The rite facilitates that the Quispillactan and the new being (seed) approach one another. The goal of incorporating the new seed variety into the family's plot is to diversify the crops. Incorporating the seed in Quispillacta involves two phases: (1) acquiring the new seed; and (2) the trial (*prueba*). These modalities listed and sketched in Table 8.2 represent the initial modalities of "germplasm maintenance," within the Quispillactan strategy of "*in situ*" conservation.

Table 8.1   Incorporation of Seed to the New Quispillactan Family

YEARS OF BECOMING ACCUSTOMED TO			
1st Year → 2nd Year → 3rd Year → 4th Year → 5th Year < ⊃			

	1st Agricultural Season	2nd Agricultural Season	3rd Agricultural Season	4th Agricultural Season
Approaching	GARDEN	TRIAL TO THE SOIL	TRIAL TO THE WEATHER	INCORPORATION
	Special plot	In the same zone	In different zones	In different plots

PHASES OF THE "TRIAL"

Years of understanding, caring for, teaching, and protecting each other

Source: Adapted from Machaca (1993).

Table 8.2   The Most Important "Visible" Forms or Modalities of Approaching the Seed

DURING THE HARVEST:

Modality:	- *Hurquchakuykuy* (To separate silently without the owner's awareness)
Actors:	- *Allaq* (3 to 5 men, heads of family, open the furrows and uncover the tubers). Search for *wanllas* (the biggest potatoes)
	- *Pallaq* (in general women and children pick up the tubers)

AFTER THE HARVEST:

Modality:	- *Maskapa* (re-search of tubers) - *Pallapa* (re-search of grains)
Actors:	- *Pallaq*

OTHER MODALITIES:

- Beyond the family circle or groups of collective work within the *ayllu*: Communal Assemblies.
- *Ruykay* (barter/*trueque*)
- *Haymay* (to help later)
- *Yanapakuy* (To cooperate)
- *Llankin* ("gift")

Source: Adapted from Machaca (1993).

Acquiring the seed

Harvest time presents a unique opportunity for the Quispillactan to approach the seed. More than two families participate in the harvest of tubers through collective and reciprocal work arrangements. Collective work provides an opportunity for a farmer to "approach" new tuber varieties in neighboring fields. The farmer takes note of characteristics such as color, number of tubers produced per plant, tuber size or unique appearance (ABA 1993; Machaca 1993). According to tradition, seeds of desirable tubers are collected surreptitiously — taken from one family and welcomed into another; considered a member of the family, the seed would "resent" being freely exchanged. The new "owner" of the seed must be especially attentive to the new tuber to facilitate its adaptation to the family.

Phases of the "trial"

The "trial" begins with sowing the seed in designated fields, such as the household garden (see Table 8.1) During the first 2 years of the trial, the goal is to "live together" with the seed and to evaluate the phenological and phenotypical characteristics of the new variety. "The home gardens constitute centers of [Quispillactan] experimentation, for both adaptation to the soil and climate, as well as other elements of [Quispillacta's] agriculture such as the *ratay* — the development of congeniality between plant and farmer" (Machaca 1993:174, author's translation). Rather than seeking to identify an ideal-type with superior characteristics, the farmer evaluates the plant's "acceptance" of the family and the field (Machaca 1993).

After the second year of the trial, the seed either rejects the new "way of living" or it accepts the new family. Because the Quispillactan provides the seed with special conditions, such as good soil, adequate sunlight and water, and dialogue during the first two years of the trial, accepting the family does not mean full incorporation of the seed. During the third year, the farmer sows the seed in different fields on the same ecological floor to test for favorable soil conditions (Machaca 1993). In the fourth year of the trial, the Quispillactan plants seed in different ecological floors to evaluate the plant's climatic range. As a result of these stages, the invited seed may lose its "way of being" due to changes in soil climatic conditions.

The fifth year is an advanced stage in the series of trials: incorporation. There is never full incorporation of the seed. "Seeds, like 'persons' have a biological cycle.... The biological cycle of the seeds, being subject to activity and fatigue, implies then the withdrawal from its original *ayllu*: a way of asking for a just rest" (Machaca 1993:175–176, author's translation).

The Asociación Bartolomé Aripaylla

With over 900 non-governmental organizations in Peru today, the Asociación Bartolomé Aripaylla (ABA) from Ayacucho is a unique local non-governmental organization: its members are indigenous peoples of Quispillacta. It did not take many years for ABA members to realize that in order to work within their community they had to reside in their community. In a short time, because of their insertion into this locality as part of the Quispillacta community "as another *ayllu*," ABA has been able to tune in to the Andean world, to their local *pacha*.

Nine years ago, ABA initiated a project of recovery of agricultural technologies and Andean crops within Quispillacta, under the central premise of strengthening Andean culture in their region. Seed fairs are part of a broader strategy of "Rescue and gathering of local and regional germplasm." Within the framework of this strategy, activities include (1) cataloging technological practices; (2) collection and cataloging of local germplasm; (3) gathering of regional germplasm; (4) communal and group sowing-plots of germplasm; (5) exchange of knowledge and seeds; and (6) devising measures to prevent genetic erosion (Machaca 1993). Between 1991 and 1993, three seed fairs were organized: The First Exhibition of Andean Germplasm in Quispillacta (1991), the Exhibition of Traditional Seeds and Crafts in Ayacucho City (1992), and the Second Fair–Exhibition of Andean Seeds in Quispillacta (1993). A brief presentation of the first two fairs will illustrate how seed fairs promote crop diversity within a framework of cultural affirmation.

First Andean agricultural fair/exhibition in Quispillacta

ABA worked jointly with community officials in Quispillacta to organize the First Exhibition of Andean Germplasm. All community members, as well as the National University of San Cristobal of Huamanga at Ayacucho, were invited to participate. The fair targeted the exchange and conservation of germplasm, cataloguing existing crop diversity and variability, and the identification of Quispillactans who maintain high levels of on-farm diversity. Peasants with the greatest variability of ecotypes per crop or a demonstrated interest in sharing their knowledge about caring for plant diversity received special recognition in the form of agricultural tools. On average, the "winners" maintain 11 crops and 74 ecotypes (Table 8.3). Such rich diversity provides evidence of the innate and remarkable capacity for the Quispillacta *comuneros* to enrich the pool of seed diversity in the Andes. The participation of the university facilitated a wider exchange of knowledge about the breeding of seed diversity and generated interest in the diversity of knowledge about the culture of the seed possessed by the Quispillacta *comuneros*. Through the exhibition of its plant germplasm collection, particularly a number of "lost" ecotypes, the university attracted the attention of the farmer participants.

The members of ABA are aware that the modality used to promote the Fair–Exhibition of Andean Seeds is not necessarily part of the community's traditional activities. They remark, however, that the essence of the event is to strengthen the breeding of seeds within their culture. Historically, communities of Ayacucho and neighboring states commonly engaged in the

Table 8.3 Participants to the First Fair–Exhibition of Andean Germplasm in Quispillacta

Name of Farmer	*Barrio*	Total Crops	Total Ecotype
1. Luis Tomaylla Conde	Pirhuamarca	13	129
2. Juan Mendoza Galindo	Cuchuquesera	15	111
3. Teodosio Flores Galindo	Llaqtahuaran	12	77
4. Felipa Nuñez Cisneros	Pirhuamarca	12	73
5. Vicente Galindo Mendieta	Tuco	13	70
6. Cristina Nuñez de Conde	Cuchuquesera	10	66
7. Aquilino Galindo Conde	Llaqtahuaran	10	60
8. Perpetua Vilca Espinoza	Llaqtahuaran	10	54
9. Emiliano Casavilca Nuñez	Tuco	9	52
10. Marcelino Tomaylla Vilca	Cuchoquesera	9	51
Special Awards			
1. Pastor Galindo Ccallocunto	Puncupata	2	47
2. Otropia Ccallocunto	Local	11	68
3. Viviana Nuñez de Huaman	Cuchuquesera	10	54

Source: Machaca (1993).

exchange of seeds, linking cultural and agricultural activities (Machaca and Machaca 1994).

ABA's second exhibition of Andean seeds

The goals of the second fair were to exhibit the native seeds "bred" by Quispillactans, exchange seeds and wisdom, and to expand the nurturing of Andean seed diversity. Organizers emphasized the role of farmers who maintain great genetic variability on-farm — identified at the first seed fair. The exhibition of Andean seeds corresponded with the three major organizational units of the Quispillacta community: the family, the extended family (*ayllu*) and the collective (*barrio*). A total of 67 of 574 families participated, presenting a total of 3,134 samples of 12 species (Table 8.4). At the *ayllu* level, five groups exhibited diverse ecotypes that community members breed collectively (Tables 8.4, 8.5, and 8.6). Collectively, they presented 377 ecotypes for 11 crops (Table 8.7). Only a single *barrio* participated, presenting 21 ecotypes of potato (Table 8.7). In general there is a high number of ecotypes per species for tubers, grains, and legumes. High variability tends to concentrate in potato (*Solanum tuberosum*), olluco (*Ullucus tuberosus*), oca (*Oaxalis tuberosa*), maswa (*Tropaeolum tuberosum*), maize (*Zea mays*), and poroto (*Phaseolus vulgaris*). Variability tends to narrow in terms of tarwi (*Lupinus mutabilis*), quinua (*Chenopodium quinoa*), achita (*Amaranthus caudatus*), linaza, kañiwa (*Chenopodium pallidicaule*), and calabaza (*Lagenaria* spp.). Kañiwa (*Chenopodium pallidicaule*) — presented by two families — was not cultivated in Quispillacta until 1991. Farmers gathered the varieties in the state of Puno (Machaca and Machaca 1994: 21–22). This illustrates the continuous

Table 8.4 Total Number of Participants at the "II Fair–Exhibition of Andean Seeds per Barrio"

			Participated in the Fair	
Barrio	Total No. of Active *Comuneros*	No. *Ayllus*	No. *Comuneros*	No. Fairs
1. Llaqta	17	—	10	—
2. Socobamba	16	1	5	—
3. Llaqtahurán	57	2	13	2
4. Pirhuamarca	40	6	9	—
5. Puncupata	48	1	8	1
6. Pampamarca	70	—	7	—
7. Cuchoquesera	46	1	4	1
8. Tuco	58	—	5	1
9. Yuraq Cruz	24	—	2	—
10. Catalinayuq	81	1	2	1
11. Huertahuasi	43	1	1	—
12. Union Potrero	74	2	1	—
TOTAL	**574**	**15**	**67**	**6**

Source: Machaca and Machaca (1994).

incorporation of new seeds to the peasant's family plot and the dynamic exchange of genetic resources.

Most seed fairs in the Andes are foreign strategies to motivate the conservation of crop genetic resources within Andean indigenous communities. In the Andean region, there is still a tradition of weekly agricultural fairs that take place in most of the capitals of districts and provinces that needs to be assessed and strengthened. In addition, there are a number of annual regional fairs (Tapia and De la Torre 1993). Seed fairs, when they are part of a cultural affirmation process, can contribute to the regeneration of agrobiodiversity as well as the local *pacha* in its three basic collectivities.

Conclusion

This chapter discusses the complexity of the Culture of the Native Seed in the Peruvian Andes, particularly within the community of Quispillacta in the state of Ayacucho. The regeneration of a great diversity of ecotypes within the Andean community implies a unique way of relating to the world (cosmology), a particular way of knowing (epistemology), and a special way of being (ontology). These three levels contrast with Western development projects, in particular with agrobiodiversity conservation projects carried out by governmental and non-governmental organizations at the local, national, and international levels. Any proposal to "preserve," protect, and enrich biological diversity (*in situ* and *ex situ*) must seriously consider the ethnic and cultural diversity of indigenous peoples' communities. In the Andes, the protection and revitalization of agrobiodiversity is inseparable from

Table 8.5 Total of Samples Presented by Crop/*Barrio* at the "II Fair–Exhibition of Andean Seeds"

Barrio Exhibitor	Papa	Olluco	Oca	Maswa	Maíz	Poroto	Tarwi	Quinua	Achita	Linaza	Kañiwa	Calabaza	Total Crop
1. Localidad	47	20	12	20	128	32	—	—	—	—	—	—	266
2. Yuraq Cruz	41	09	—	—	—	—	—	—	—	—	—	—	50
3. Llaqtahurán	298	54	80	97	231	149	23	11	4	1	—	1	949
4. Huertahuasi	15	—	—	—	—	—	—	—	—	—	—	—	15
5. Pirhuamarca	145	30	44	57	116	57	4	5	2	1	—	—	461
6. Socobamba	104	16	18	23	51	5	1	3	2	1	—	2	226
7. Tuco	45	11	16	38	93	14	1	4	3	1	—	—	226
8. Unión Potrero	13	06	7	—	18	—	—	—	—	—	—	—	44
9. Puncupata	119	23	12	39	68	11	2	3	3	2	—	2	284
10. Catalinayocc	26	03	5	3	25	6	3	2	—	—	—	—	73
11. Pampamarca	40	12	—	14	100	11	—	—	2	—	—	—	178
12. Cuchoquesera	67	24	14	24	177	36	15	5	2	3	1	2	310
TOTAL	960	208	208	315	1,007	321	49	33	16	9	1	7	3,134

Source: Machaca and Machaca (1994).

Table 8.6 Peasant Farmer Winners at the "II Fair–Exhibition of Andean Seeds"

Farmer	Papa	Olluco	Oca	Maswa	Maíz	Poroto	Tarwi	Quinua	Achita	Linaza	Kañiwa	Calabaza	Total Crops	Total Ecotype
1. Dámaso Mendoza	42	14	7	11	46	20	11	4	3	1	1	1	12	161
2. Marcelino Tomaylla	27	6	9	13	33	16	3	—	—	—	—	—	7	107
3. Alejandro Galindo	36	4	5	10	22	19	2	2	—	—	—	—	8	100
4. Guillermo Vilca	22	6	7	12	28	16	4	1	2	1	—	—	10	99
5. Gerardo Nunez	34	3	7	11	23	9	6	1	1	—	—	—	10	96
6. Pastor Galindo	64	6	—	10	18	—	—	—	—	—	—	—	4	98
7. Teodosio Flores	44	5	9	4	19	14	2	—	1	—	—	—	8	98

Source: Machaca and Machaca (1994).

Table 8.7 Ecotypes Presented by Group and/or *Ayllus* in the "II Fair–Exhibition of Andean Seeds"

Grupo and/or Ayllu	Papa	Olluco	Oca	Maswa	Maíz	Poroto	Quinua	Achita	Linaza	Kañiwa	Calabaza	Total Crops	Total Ecotypes
1. Atahualpa	14	9	5	13	25	12	5	2	—	1	—	10	89
2. Rikchariyllaqta	29	8	3	7	—	—	4	—	—	—	—	5	50
3. Amaqilla	36	13	9	13	18	12	—	2	—	—	—	9	109
4. Espinoza	32	5	6	12	36	15	—	—	—	—	—	7	108
5. *Barrio de Tuco*	21	—	—	—	—	—	—	—	—	—	—	1	21
TOTAL	**132**	**35**	**23**	**79**	**79**	**39**	**9**	**4**	**—**	**1**	**—**	**10**	**377**

Source: Machaca and Machaca (1994).

Andean Quechua and Aymara agriculture. The case of the community of Quispillacta calls attention to a particular indigenous way of *"in situ"* conservation and the need to acknowledge critical underlying differences between two different systems of *in situ* conservation of agrobiological diversity.

It is worthy to note the conclusions of a group of scientists at a meeting on knowledge, innovations, and practices of indigenous and local communities:

> 1. As scientists, we recognize that this is not a purely scientific issue, but at the same time, that the involvement of scientists is critical.
> 2. The question itself has to be rephrased. The challenge is not to find the ways to integrate, in modern management practices, knowledge, innovations and practices of indigenous and local communities. Rather, it is to define, **in collaboration with indigenous and local communities**, which modern tools may be of help to them, and how these tools might be used, **to strengthen and develop their own strategy for conservation and sustainable use of biological diversity, fully respecting their intellectual and cultural integrity and their own vision of development** (UNEP 1994:4, emphasis added).

Only a handful of innovative and flexible foreign and nongovernmental organizations provide funding for cultural affirmation activities such as those of ABA. Supporting these activities presents both a challenge and a potential role for national and international cooperation agencies and bodies. An unavoidable and critical factor postponed by most states all over the world is the resolution of the Indigenous Question, that is, the struggle of indigenous peoples for self-determination, control over their territories, and resources. In the long term, this central issue will contribute to secure on-farm conservation of agro-biodiversity.

References

Acuerdo de Cartagena. 1991. Decisión 291. Régimen Común de Tratamiento a los Capitales Extranjeros y sobre Marcas, Patentes, Licencias y Regalías. *Quincuagésimo Período de Sesiones Ordinarias de la Comisión*. Lima, Peru: 21–22 de Marzo.

Adams, R.N. 1994. Introduction. Ethnic Conflict and Governance in Comparative Perspective. *Working Paper Series*. Number 215, The Latin American Program. Woodrow Wilson International Center for Scholars.

Alcorn, J.B. 1991. Ethics, economies and conservation. In *Biodiversity: Culture, Conservation, and Ecodevelopment*, M.L. Oldfield and J.B. Alcorn (eds.). Boulder, CO: Westview Press.

Asociación Bartolomé Aripaylla (ABA). 1993. Diversificación de Germoplasma Agrícola en Quispillacta. Año Agrícola 1991–1992. Asociación Bartolomé Aripaylla de Ayacucho. Manuscript.

Ballón, E., R. Cerrón-Palomino, and E. Chambi. 1992. *Vocabulario Razonado de la Actividad Agraria Andina. Terminología agraria quechua.* Centro de Estudios Regionales Andinos "Bartolomé de Las Casas."

Browning, J.A. 1991. Conserving crop plant-pathogen coevolutionary processes *in situ.* In *Biodiversity: Culture, Conservation, and Ecodevelopment,* M.L. Oldfield and J.B. Alcorn (eds.). Boulder, CO: Westview Press.

Buttel, F.H. 1993. Twentieth Century Agricultural-Environmental Transitions: A Preliminary Analysis. Discussion Draft. Paper prepared for presentation at the Agrarian Studies Seminar, Yale University.

Chambi, N. and W. Chambi. 1995. *Ayllu y Papas. Cosmovisión, religiosidad y agricultura en Conima, Puno.* Asociación Chuyma de Apoyo Rural "Chuyma Aru."

Colchester, M. 1994. Salvaging Nature. Indigenous Peoples, Protected Areas and Biodiversity Conservation. *United Nations Research Institute for Social Development, DP 55.*

Descola, P. 1989. *La Selva Culta.* Quito: Abya Ayala.

Durning, A.T. 1992. *Guardians of the Land: Indigenous Peoples and the Health of the Earth.* Washington, D.C.: Worldwatch Institute.

Gonzales, T. 1987. The Political Economy of Agricultural Research and Education in Peru, 1902–1980. Master's thesis, University of Wisconsin, Madison.

Gonzales, T. 1996. Political Ecology of Peasantry, the Seed, and NGOs in Latin America: A Study of Mexico and Peru, 1940–1996. Ph.D. thesis. University of Wisconsin, Madison.

Gonzales, T., N. Chambi, and M. Machaca. In press. Agricultures and cosmovision in contemporary Andes. In *Cultural and Spiritual Values of Biodiversity,* Darrell Posey (ed.). Cambridge University Press.

Gonzalez, E. 1986. *Economia de la Comunidad Campesina.* Lima: Instituto de Estudios Peruanos.

Greslou, F. 1991. La Organización Campesina Andina. In *Cultura Andina Agrocentrica,* PRATEC (ed.). Lima: PRATEC.

Griffin, K. 1974. *The Political Economy of Agrarian Change. An Essay on the Green Revolution.* London: The Macmillan Press Ltd.

Grillo, E. 1993. Afirmación Cultural en los Andes. Draft.

Haynes, R.P. 1985. Agriculture in the U.S.: Its impact on ethnic and minority groups, *Agriculture and Human Values* 3:1–3.

ISNAR. 1987. *El Modelo de Investigación, Extensión y Educación en el Perú.* Estudio de Un Caso, Vols I, III, and IV. International Service for National Agricultural Research ISNAR R30s.

Jennings, B.H. 1988. *Foundations of International Agricultural Research. Science and Politics in Mexican Agriculture.* Boulder, CO: Westview.

Juma, C. 1989. *The Gene Hunters: Biotechnology and the Scramble for Seeds.* Princeton, NJ: Princeton University Press.

Kloppenburg, J. 1988. *First the Seed.* Cambridge: Cambridge University Press.

Latour, B. 1993. *We Have Never Been Modern.* Cambridge, MA: Harvard University Press.

Lumbreras, L.G. 1992. Cultura, tecnologia y modelos alternativos de desarrollo, *Comercio Exterior* 42:199–205.

Machaca, M.M. 1993. Actividades de crianza de semillas en la Comunidad Campesina de Quispillacta, Ayacucho, acompañadas por la Asociación Bartolomé Aripaylla. Campaña 1992–93. In *Afirmación Cultural Andina.* Peru: Proyecto Andino de Tecnologías Campesinas, PRATEC.

Machaca, M.M. and M. Machaca. 1994. *Crianza Andina de la Chacra en Quispillacta. Semillas — Plagas y Emfermedades.* Peru: Asociación Bartolomé Aripaylla.

Matos Mar, J. 1991. "Los pueblos Indios de Iberoamerica." *Pensamiento Iberoamericano,* N. Maxted, B.V. Ford-Lloyd, and J.G. Hawkes (eds.). 1997. *Plant Genetic Conservation: The* in situ *Approach.* London: Chapman & Hall.

Mayer, E. and O. Dancourt. 1994. Recursos naturales, medio ambiente, tecnología y desarrollo. In *Perú: El Problema Agrario en Debate, SEPIA V,* C. Menge (ed.), 479–533. Peru: SEPIA-CAPRODA.

Nabhan, G. 1985. Native American crop diversity, genetic resource conservation, and the policy of neglect, *Agriculture and Human Values* 3:14–17.

National Research Council. 1989. *Alternative Agriculture.* Washington, D.C.: National Academy Press.

Oasa, E.K. and B. Jennings. 1982. Science and authority in international agricultural research, *Bulletin of Concerned Asian Scholars* 14:30–44.

Oldfield, M.L. and J.B. Alcorn (eds.). 1991. *Biodiversity: Culture, Conservation, and Ecodevelopment.* Boulder, CO: Westview Press.

Pearse, A. 1980. *Seeds of Plenty, Seeds of Want.* New York: Oxford University Press.

Pimbert, M. 1994. The need for another research paradigm, *Seedling* 11:20–32.

Redclift, M. 1987. *Sustainable Development. Exploring the Contradictions.* London and New York: Methuen & Co.

Richards, P. 1993. Cultivation: knowledge or performance? In *Anthropology of Development,* Mark Hobart (ed.). London and New York: Routledge.

Stavenhagen, R. 1990. *The Ethnic Question.* Tokyo: United Nations University Press.

Tapia, M.E. and De la Torre, A. 1993. La Mujer Campesina y las Semillas Andinas. Lima, Peru: UNICEF-FAO.

Toledo, V.M. 1990. The ecological rationality of peasant production. In *Agroecology and Small Farm Development,* M. Altieri and S. Hecht (eds.). Boca Raton, FL: CRC Press.

UNEP. 1994. Open-ended Intergovernmental Meeting of Scientific Diversity on Biological Diversity, Annexes to the Draft Report of Subcommittee. *UNEP/CBD/IGSc/1/ SC.II/L.2/Add.2* Mexico City, II 11–15, April.

Van der Ploeg, J.D. 1993. Potatoes and knowledge. In *An Anthropological Critique of Development. The Growth of Ignorance,* M. Hobart (ed.). London and New York: Routledge.

Van Kessel, J. and D. Condori. 1992. *Criar la Vida: Trabajo y Tecnología en el Mundo Andino.* Santiago, Chile: Vivarium.

WRI, IUCN, UNEP. 1992. *Global Biodiversity Strategy. Guidelines for Action to Save, Study, and Use Earth's Biotic Wealth Sustainably and Equitably.* World Resources Institute, WRI, The World Conservation Union, IUCN, United Nations Environment Programme, UNEP.

Zandstra, I. 1993. Director of Peruvian plant genetic resources program faces many obstacles, *Diversity* 9:1–2, 29–31.

chapter nine

On-farm conservation of crop diversity: policy and institutional lessons from Zimbabwe[1]

Elizabeth Cromwell and Saskia van Oosterhout

Introduction

The milestone international Convention on Biological Diversity (CBD), which was signed at the United Nations Conference on the Environment and Development in 1992, emphasizes in Article 8 that conservation of agricultural biodiversity is important in farmers' fields as well as in protected areas and in gene banks. It states that signatory countries should:

> regulate or manage biological resources important for the conservation of biological diversity **whether within or outside protected areas**, with a view to ensuring their conservation and sustainable use [Article 8 (c)]; and
> respect, preserve and maintain knowledge, innovations and practices of indigenous and local communities embodying traditional lifestyles relevant for the conservation and sustainable use of biological diversity **and promote their wider application**... [Article 8 (j)] (UNEP 1994:8–9, emphases added).

However, this approach to conserving agricultural biodiversity remains unfamiliar, ambiguous, and controversial to many people. In particular, there has been little exploration of the economic, sociocultural, and environmental variables influencing farmers' attitudes toward maintaining crop diversity on-farm, and therefore little understanding of farmers' willingness to get involved in on-farm conservation.

This chapter begins with a presentation of the results from recent research into these issues. We then use these results to explore how farmers need to be supported if they are to maintain on-farm crop diversity. We conclude by offering some insights into the viability of using on-farm conservation as a tool for conserving agricultural plant genetic diversity, based on these results. We use evidence from the Southern Africa region because this is one region of the world where the challenge to undertake on-farm conservation has been taken up actively, with a number of important initiatives at the regional, national, and farm levels, as well as an active and vibrant debate among the main stakeholders.

The evidence that we present is taken from the work of the Sorghum Landrace Study of the Government of Zimbabwe Department of Research and Specialist Services, and of the Darwin Initiative for *In Situ* Conservation in Zimbabwe, which was implemented by the Overseas Development Institute and the Sorghum Landrace Study.[2] We have also benefited from the thoughtful insights into crop diversity issues offered by participants at a workshop on Supporting Diversity Through Sustainable Livelihoods: What Are Farmers' Choices? held in Harare in November 1996 under the auspices of the Darwin Initiative for *In Situ* Conservation in Zimbabwe.[3]

Definitions

The CBD states that biodiversity "means the variability among living organisms from all sources and the ecological complexes of which they are part; this includes diversity within species, between species and of ecosystems" (UNEP 1994:4). For our research, we focused on two of these measures in relation to agricultural biodiversity: diversity within species (varieties/landraces) and between species (crops).

Landraces and varieties produced by the formal sector are of course different in important respects. We define a landrace as a local farmer's variety of a particular crop. Landraces exhibit varying degrees of morphological and genetic integrity and may change with time, but they are recognized by farmers on the basis of a number of morphological and agronomic criteria. By formal sector varieties ("modern" varieties), we mean distinct and stabilized assemblages of local or exotic material which have been selected for certain criteria, most often higher yield and pest or disease resistance, by formally qualified plant breeders. In this research, we use "variety" to refer to both types of material, except where specifically stated otherwise.

As regards crops, we focus on the on-farm conservation of "small grains": sorghum (*Sorghum bicolor*), pearl millet (*Pennisetum vulgare*), and finger millet (*Eleusine coracana*). We chose this focus because these are crops for which the Southern Africa region is an important center of diversity but which, at the same time, are under great threat from the "modern" cash-based economy which promotes maize monoculture. Note that reference to "maize cropping" throughout this chapter refers specifically to the cultivation of commercially sold hybrid varieties, not to the growing of open-pollinated varieties, which is much less common in Zimbabwe.

Even with these relatively tight definitions, it is difficult to know how to measure diversity in farmers' fields. We decided to use three very simple measurements of on-farm crop diversity:

1. the number of crops grown on-farm;
2. the number of crop varieties grown on-farm; and
3. the proportion of the total farm area allocated to growing small grains.

We selected these criteria based on the assumption that there is a direct relationship between the number of crops and crop varieties grown on-farm and the level of crop diversity on-farm. Likewise, we assumed that the greater the proportion of the farm area allocated to growing small grains, rather than hybrid maize, the greater the likelihood of a high level of on-farm diversity, compared to those farms with smaller portions of land allocated to growing small grains.

It is difficult to draw conclusions about whether certain crops and varieties are more "valuable" or "important" than others; or whether landrace material is preferable to "modern varieties." We have not attempted to do this, i.e., to assess "optimal" levels of diversity within a given farming system; rather, as we explain in the section on methodology below, our aim is simply to identify those economic, sociocultural, and environmental variables that influence whether diversity on one farm is higher or lower than on another.

The study area[4]

The research on which this chapter is based was carried out in Mutoko and Mudzi districts in Zimbabwe, which lie next to each other northeast of the capital city Harare and straddle one of the main roads to the Mozambique border (see Figure 9.1). These two districts were chosen because they encapsulate the wider situation in the Southern Africa region: while the districts are rich in crop diversity, farmers' ability to maintain this diversity on-farm is apparently threatened by intense livelihood pressures.

Mean annual rainfall is less than 600 mm over most of the two districts,[5] with a high likelihood of severe midseason dry spells during the rainy season, and of droughts occurring every 3 to 4 years. Population densities range from nearly 50 people per sq km in Mutoko to around 30 people per sq km in the more easterly and remote Mudzi. There has been considerable economic and sociocultural dislocation in the study area, caused by two primary factors. First, as a consequence of the Independence struggle, the then government of Rhodesia forcibly relocated many families into camps

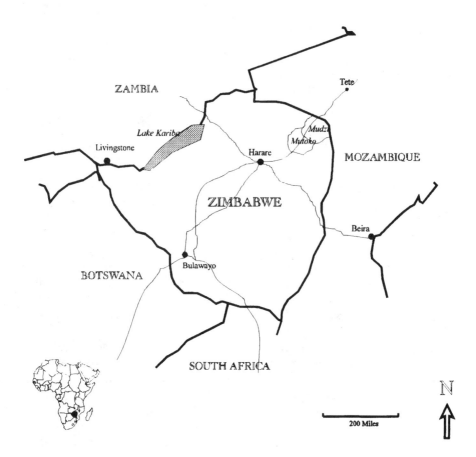

Figure 9.1 Location of Mutoko and Mudzi districts.

between 1976 and 1979. Second, the Mozambican civil war spilled over into the border areas of Zimbabwe, so as late as 1993 families in some parts of the area were sleeping in village schoolrooms for safety. This dislocation has had a disruptive effect on many aspects of traditional farming systems.

The average household consists of five people, but many able-bodied men work away from home for all or part of the year, resulting in a high dependency ratio. Approximately 60% of the people actively farming the land are women, although only just over 20% of households are formally or informally headed by women. Most household heads have some education, usually at the primary level. Small grains are important crops in the area because of their drought tolerance, but over time maize has become an increasingly dominant part of the cropping system for reasons that will be discussed later. Land is allocated to families by traditional authorities (chiefs) and mean holding size is 2.5 ha, although this varies widely.

At present, land availability is not usually a limiting factor in crop production. Rather, shortage of labor, infertile soils, and lack of draught power are more significant. Over half of all households apply fertilizer, primarily inorganic; the use of organic fertilizers (leaf litter and cattle manure) is very limited. Crop production is insufficient to produce a surplus at the household level for most families, who are therefore net buyers of grain and reliant on non-agricultural activities, including casual labor, informal gold-panning, or craft-making, for a significant portion of total household income. In addition, 30% of household heads work away from home and 75% of households receive financial help from urban relatives.

In general terms, Mutoko is more economically developed than Mudzi, located closer to jobs and markets in Harare, and better served by transport and other economic infrastructure. Mutoko has more "Master Farmers"[6] and members of farmer groups, as well as more educated household heads. Cropping systems are more diversified and farmers in Mutoko employ more casual labor for on-farm activities. Soils are less fertile, however, as they have been continuously cropped over a longer period of time, and hence more inorganic fertilizer is used in Mutoko than in Mudzi, where farmers generally still consider their soils to be fertile.

Methodology

The purpose of our research was to test the following hypothesis:

> *H: farmers may not be willing to maintain crop diversity on-farm due to the influence of exogenous economic, sociocultural, or environmental variables.*

Our assumption was that our three chosen measures of on-farm crop diversity could be taken as proxy indicators of farmers' willingness to maintain crop diversity on-farm, i.e., greater numbers of crops and varieties on-farm and larger farm areas allocated to small grains demonstrate a greater willingness to maintain crop diversity on-farm. Our challenge, therefore, was to identify which economic, sociocultural, and environmental variables are strongly correlated — positively or negatively — with our three chosen measures of on-farm crop diversity.

Our research took place over the course of the 1995–1996 cropping season. Twelve villages in Mutoko and Mudzi districts were selected to represent a range of different economic, sociocultural, and environmental conditions. The research started with participatory rural appraisal exercises with representative groups of farmers in each village. Our aim was to gain a thorough understanding of farmers' thought systems concerning on-farm crop diversity. Accordingly, we used seven different exercises: mind-mapping; history time-lines; wealth ranking; social mapping; matrix ranking of farmers' problems; mobility mapping; income and expenditure ranking; and semi-structured interviews on seed sourcing.

The results from the participatory rural appraisal exercises helped us to draw up a questionnaire about on-farm crop diversity, which was applied to 25 farm families in each of the 12 villages, i.e., to 300 farm families altogether. Together with relevant background information, the questionnaire provided data for each family for the range of economic, sociocultural, and environmental variables that we believed might influence levels of on-farm crop diversity. These are described in Box 9.1. Our decisions about which variables to include in the questionnaire were based on our understanding of on-farm crop diversity obtained from the wider literature, previous research carried out in Mutoko and Mudzi by the Sorghum Landrace Study, and participatory rural appraisal results. Multiple regression analysis was then carried out on the data from the questionnaires, with the aim of identifying which specific economic, sociocultural, and environmental variables appear to be strongly associated with high levels of on-farm crop diversity, as measured by our three proxy indicators. Box 2 specifies the equations used in each case. The results from the multiple regression analysis are discussed in the next section.

Box 9.1 Variables Hypothesized to Influence On-Farm Crop Diversity in Zimbabwe*

Economic variables

Area cultivated (ha) (e1)

Sum of areas planted to each crop in current season (i.e., does not include fallow areas).

Wealth status (e2)

During the participatory rural appraisal exercise on wealth ranking, farmer groups assigned families as being 'rich,' 'average,' or 'poor' according to a wealth index that took account of numbers of livestock; type of housing; type of farm implements; size of crop production; employment of labor; employment of household members in town; and children's education. Interestingly, farm size was not considered to be a relevant indicator of wealth, on account of the fact that some families had relatively large holdings to compensate for the poor quality of the land.

Shortage of labor

We hypothesized that two dimensions of labor shortage might be influential:

☐ casual labor (e3) – whether respondents or other family members worked as casual laborers for other farmers, *and* considered this to cause them to delay or neglect their own farming duties;

☐ labor for seed sourcing (e4) – whether respondents considered they do not normally have enough labor for seed sourcing.

Maize-mindedness

(note: the phenomenon of maize-mindedness is explained in more detail in Economic variables section in text.

Two dimensions of maize-mindedness were hypothesized to be potentially influential:

☐ increased maize area over time (e5): whether the proportion of the farm area planted to maize in the current season is greater than at Independence (1980);

☐ proportion of cultivated area planted to maize (e6): area planted to maize in the current season as a proportion of total area cultivated.

Extension contact (e7)

Whether respondents consider they have contact with the extension services for the purpose of receiving extension advice (as opposed to for free inputs, drought relief, etc.).

Seed security

We hypothesized that two aspects of seed security might be influential:

☐ access to preferred varieties (e8): whether respondents are growing as many varieties as they would like;

☐ secure access to seed (e9): whether respondents had sufficient quantity of seed in the current *and* previous season (the latter being a major drought year).

Location (e10)

In either Mutoko or Mudzi District.

Sociocultural variables

Age of household head (s1)

Local position of authority (s2)

Respondents were asked whether they or the head of the household held any position of authority in the local community, either elected (e.g., a local councillor) or traditional (e.g., chief, traditional healer).

Education of household head (s3)

Respondents were asked whether the household head had any education (yes/no).

Sex of household head (s4)

Cropping decisions made by women family members (s5)

Respondents were asked who in the household makes the decisions about which crops and varieties to grow, and the area to allocate to each crop.

Value placed on small grains by family (s6)

Respondents were asked to identify and rank various agronomic, economic, and cultural reasons why growing small grains is important to them (e.g., for disease resistance, for food security, for the spirits)

Environmental variables

On-farm environmental variability (v1)

Respondents were asked to rank the degree of variation in on-farm slope, soil type, and other aspects of terrain.

On-farm environmental quality (v2)

Respondents were asked to state whether pests, poor soils, and leaching of nutrients were problems on their farm.

Access to on-farm resources (v3)

Respondents were asked whether they had access to a range of resources on-farm, including fruit trees, agro-forestry trees, a variety of good soils, and good water supplies.

Access to off-farm resources (v4)

Respondents were asked whether they had access to a range of resources off-farm, such as wild fruit trees, leaf litter, forest area, thatching grass, or grazing.

---

* Note that the characterization of variables as "economic," "sociocultural," or "environmental" is based on our own best judgments and is slightly arbitrary in some cases.

Box 9.2   Multiple Regression Analysis Used to Investigate Variables Affecting On-Farm Crop Diversity in Zimbabwe: Specification of Equations

---

$D1 = e1 + e2 + e4 + e5 + e7 + e8 + e9 + e10 + s1 + s2 + s3+s4 +s6 + v2 +v3 + v4$
$D2 = e1 + e2 + e4 + e5 + e7 + e8 + e9 + e10 + s1 + s2 + s3+s4 +s6 + v2 +v3 + v4$
$D3 = e1 + e2 + e3 + e4 + e5 + e6 + e7 + e8 + e10 + s1 + s5 + v1 + v3 + v4$

where:

$D1$ = number of crops grown on-farm
$D2$ = number of varieties grown on farm
$D3$ = proportion of farm area allocated to small grains (measured as proportion of farm area allocated to all non-hybrid maize cereals)

and

$e1$ = area cultivated
$e2$ = wealth status
$e3$ = shortage of labor: casual labor
$e4$ = shortage of labor: seed sourcing
$e5$ = maize-minded: increased maize area over time
$e6$ = maize-minded: proportion of cultivated area planted to maize
$e7$ = extension contact
$e8$ = seed security: access to preferred varieties
$e9$ = seed security: secure access to seed
$e10$ = location

$s1$ = age of household head
$s2$ = family in local position of authority
$s3$ = education of household head
$s4$ = sex of household head
$s5$ = cropping decisions made by women family members
$s6$ = values placed on small grains

$v1$ = on-farm environmental variability
$v2$ = on-farm environmental quality
$v3$ = access to on-farm rescues
$v4$ = access to off-farm resources

Analysis

A summary of the multiple regression results on which the analysis is based is presented in Table 9.1.

Table 9.1 Variables Found to Be Related to On-Farm Crop Diversity in Zimbabwe: Summary Results of Multiple Regression Analysis

Variables Related to Number of Crops Grown On-Farm		
Independent Variable	Significance	T-value
Adjusted r^2 = 0.85680		
Mudzi District (e10)	.000	(9.180)
Access to preferred varieties (e8)	.000	5.172
Area cultivated (e1)	.000	5.039
Shortage of labor for seed sourcing (e4)	.001	3.555
Poor on-farm environment (v2)	.006	(3.036)
Secure access to seed (e9)	.012	2.716
Contact with extension (e7)	.047	2.111
Small grains valued (s6)	.080	1.835

Variables Related to Number of Crop Varieties Grown On-Farm		
Independent Variable	Significance	T-value
Adjusted r^2 = 0.80274		
Mudzi District (e10)	.000	(6.709)
Area cultivated (e1)	.000	5.191
Shortage of labor for seed sourcing (e4)	.000	3.911
Contact with extension (e7)	.002	3.426
Access to preferred varieties (e8)	.008	2.896
Small grains valued (s6)	.035	2.245
Secure access to seed (e9)	.040	2.183
Position of authority (s2)	.046	2.116

Variables Related to Proportion of Farm Area Allocated to Small Grains		
Independent Variable	Significance	T-value
Adjusted r^2 = 0.51046		
Proportion of cultivated area planted to maize (e6)	.0000	(13.366)
Poor family (e2)	.0030	2.997
Mudzi District (e10)	.0170	2.402
Age of household head (s1)	.0377	2.089
Access to off-farm resources (v3)	.0481	(1.987)
Increased maize area over time (e5)	.0609	(1.882)
On-farm environmental variation (v1)	.0789	1.764
Access to on-farm resources (v3)	.0877	1.714

Numbers of crops and crop varieties grown on-farm

Our regression analyses showed that there is a significant relationship between the number of crops and crop varieties grown on-farm and the following variables:

- *Location:* families in Mutoko grow more crops and varieties than those in Mudzi.
- *Cultivated area:* families with a larger cultivated area grow a larger number of crops and varieties than do families cultivating smaller areas.
- *Seed security:* those households that are seed secure grow more crops and varieties than those households that are not seed secure.
- *Extension contact:* those families who are in contact with the extension services grow more crops and varieties than those families who have no or minimal contact.
- *Small grains valued:* those families who value small grains highly, grow more crops and varieties than those families who do not.

In addition, the number of crops grown and the number of crop varieties grown were each significantly influenced by one of the following variables:

- *On-farm environmental quality:* those families with a good environment on-farm cultivate more crops than those with a poor on-farm environment.
- *Position of authority:* those households with a position of authority in the local community grow more varieties than those without a position of authority.

Proportion of farm area allocated to small grains

The regression analysis identified a number of variables influencing the proportion of the farm area allocated to small grains. Interestingly, only one of these (namely, location) is the same as the variables influencing how many crops and crop varieties are grown on-farm:

- *Proportion of farm area allocated to maize:* this is the variable that most strongly influences the proportion of farm area allocated to small grains. Not surprisingly, those households that allocate a smaller proportion of their farm area to maize allocate a greater proportion to small grains.
- *Wealth:* poorer families allocate a greater proportion of their farm area to small grains than rich families.
- *Location:* families in Mudzi allocate a greater proportion of their farm area to small grains than those in Mutoko.
- *Age of household head:* those families with older household heads allocate a greater proportion of their farm area to small grains than those households with younger household heads.
- *On-farm environmental variation:* those families with farms with great environmental variation allocate a greater proportion of their farm area to small grains than those with more uniform land.

- *On-farm resources*: those families with access to a large number of resources on-farm allocate a greater proportion of their farm area to small grains than those with poor access to on-farm resources.
- *Off-farm resources*: those families with poor access to off-farm resources allocate a greater proportion of their farm area to small grains than those with good access to off-farm resources.

Discussion of regression results

Perhaps one of the most important findings from the above analysis is that it is not one single set of variables, whether economic, sociocultural, or environmental (see Box 9.1 for definitions and categorizations) that determines on-farm crop diversity, but rather a complex combination of these sets of variables. As we shall see below, the complexity of this combination produces conflicting signals when trying to identify particular sets of conditions that need to be satisfied in order for farmers to be willing to undertake on-farm conservation.

Economic variables

The regression results for some economic variables influencing on-farm crop diversity are difficult to interpret. As regards area cultivated, the results suggest that families with larger cultivated areas grow a greater total number of crops and crop varieties. This is contrary to the apparently widespread assumption that families with larger areas under cultivation are less interested in on-farm crop diversity and more oriented toward monoculture.[7] At the same time, the results suggest that poorer families allocate a greater proportion of their farm area to small grains, although they grow fewer crops and varieties than do richer families. Taken together, these results imply that, although the proportion of cultivated area allocated to small grains by richer families may be proportionally smaller than that of poorer families, the absolute number of varieties will be relatively larger. This may also be a result of the fact, identified during the participatory rural appraisal exercises, that families with smaller cultivated areas consider it unwise to grow many different crop varieties: where cultivated area is limited, families prefer to concentrate on growing a few varieties.

Furthermore, other data collected via the farm family questionnaires show that poorer families are caught in a vicious circle that pushes them away from their farms in an effort to earn their living. These data show that poorer families often have to neglect important farming duties, such as planting and weeding, because they are trying to earn cash or get food off-farm for their immediate needs. Often this involves laboring for richer families at precisely the time their own farms should be planted or weeded. Thus, their harvests are poor and the following season these families are even more dependent on alternative sources of survival, their own farm is further neglected, and the vicious circle continues.

As regards the location, the regression results suggest that households in Mutoko (i.e., in locations with a higher level of "economic development") grow a greater number of crops and crop varieties. However, in Mudzi (i.e., in locations with a lower level of "economic development") households allocate a greater proportion of farm area to small grains. This may be explained in part by the fact that in recent years a non-governmental project has been distributing seed in Mutoko brought in from other areas, so some families in Mutoko district have had greater access to seed of a number of varieties, and also greater exposure to publicity concerning the value of on-farm crop diversity. Perhaps this has encouraged families in Mutoko to maintain a greater number of crops and varieties on their farms; this requires further investigation.

In discussing the influence of economic variables on on-farm crop diversity, it is necessary to explain the phenomenon which farmers call "maize-mindedness." With the attainment of Independence in Zimbabwe in 1980, the focus of agricultural research and extension turned to the so-called communal (small farm) areas to redress previous neglect, with the aim of increasing production and marketed surplus from these areas. Packages of hybrid maize seed and fertilizer started to be distributed widely to smallholder farmers and 90% of the total short-term loans handed out in the 1980s by the Agricultural Finance Cooperation, which provides credit facilities to smallholder farmers, were related to maize production (MLARR 1990). A series of droughts resulted in the further free distribution of these seed and fertilizer packages for "drought relief" in subsequent years. At the same time, market prices were adjusted to encourage greater maize production and sales.

According to our participatory rural appraisal results, all this resulted in farmers becoming increasingly oriented toward investing all the best household resources of labor, land, and agricultural inputs into the production of hybrid maize, i.e., becoming "maize-minded" as they describe it. The amount and quality of land, labor, and inputs devoted to other crops are thus primarily allocated *after* decisions concerning how and where to grow the hybrid maize crop have been made. Farmers say that "maize-mindedness" has affected on-farm crop diversity by reducing the area farmers allocate to small grains, and by reducing the number of different small grain varieties as these have become redundant in the modified farming system.

"Maize-mindedness" has not, however, translated into increased household food security. Page and Chonyera (1994) report that most maize sales — even in high potential areas — can be accounted for as "distress sales," whereby families have to sell most of their harvest in order to repay the credit received at the start of the season. This has left a large proportion of farm families food insecure and, over the years since Independence, "maize-mindedness" has resulted in families becoming severely indebted — especially in the more resource-poor, low rainfall, marginal areas such as Mutoko and Mudzi. Much circumstantial evidence is available which positively relates food security to crop diversity, but little concrete information is as yet accessible (Guveya 1996). Finally, the regression results show that household seed security, in terms of access to seed in general and access to seed of preferred varieties, has a significant influence on the number of crops and crop varieties grown. Regarding household contact with the extension services, the regression results refute the commonly held assumption that *greater* contact with extension agents results in farm families being more oriented toward monoculture and *less* interested in maintaining a large number of crops and varieties on-farm.

Sociocultural variables

According to the regression results, some sociocultural variables have a significant positive influence on on-farm crop diversity, in line with prevailing assumptions. These variables are, namely: the extent that small grains are valued within the farm family; whether the household head is relatively old; and whether the family has a position of authority within local society. We suggest that the positive correlation between the age of the household head and the proportion of the farm area allocated to small grains may pose a threat to the longer-term maintenance of on-farm crop diversity. As the older generation dies, and economic pressures on younger families continue to increase, the area allocated to small grains may become reduced to such an extent that it may be insufficient to maintain diversity at biologically meaningful and economically satisfactory levels.

Interestingly, the regression results suggest that two sociocultural variables commonly assumed to be significantly positively correlated with on-farm crop diversity do not have this effect in Mutoko and Mudzi districts. According to our regression results, the sex of the person within the household who decides which crops and varieties to plant and the area to allocate to each crop has no significant influence on any of our measures of on-farm crop diversity, contrary to the findings of, for example, Sperling and Loevinsohn (1993) and Prain and Piniero (1994). Likewise, whether or not the household head is educated had no significant bearing on crop diversity. As regards the former, this may be because, although women farmers place great importance on the nutritional value and storage quality of small grains, the great amount of labor associated with the growing of small grains in terms of thinning, bird scaring, threshing, dehulling, and pounding has become the domain of woman, as more children now attend school, so workloads for women farmers have increased. Women are therefore less keen to grow large areas of small grains than they were at one time.

Environmental variables

The regression results suggest that environmental variables affect on-farm crop diversity more by influencing area allocation decisions than by affecting decisions about the number of crops and varieties grown. As regards the on-farm environment, the results appear to show that the quality of the on-farm environment positively affects the number of crops grown, and the diversity of the on-farm environment positively affects the proportion of the cultivated area allocated to small grains. This is probably because on-farm environmental diversity presents families with micro-niches which can be exploited by growing a diverse array of crop varieties. The regression results imply that access to on-farm resources positively affects the proportion of the farm allocated to small grains, but families who have greater access to off-farm resources tend to allocate less land to small grains.

Implications for on-farm conservation of crop diversity

In this section, we present our interpretation of the implications of the above analysis for on-farm conservation projects and programs. First, we present what we believe the results tell us concerning how to identify farm families who are willing to maintain on-farm crop diversity; second, we present what we believe the results tell us regarding how these pro-diversity families can be supported in their efforts to maintain on-farm crop diversity.

Identifying farm families willing to maintain on-farm crop diversity

The usual priority for on-farm conservation projects and programs is to identify farmers who are already growing a relatively large number of crops and crop varieties (Maxted et al. 1997; Maxted et al. in press). Our results suggest that projects and programs wishing to do this should target households with larger farms and a good on-farm environment (meaning few pests and fertile soils), who have secure sources of seed, who feel they have good extension contact, who value small grains highly, and who are headed by someone with a position of authority within the local community.

Having identified these households, projects and programs usually then want to find out which households within this group are more likely to allocate a large proportion of their farm area to small grains. Our results suggest that these will be poorer households headed by an older person. Resource-wise, they will be households with good access to on-farm resources, and their farms will show considerable environmental variation, but off-farm resources will not be of great importance.

Supporting pro-diversity farm families

Our results suggest that families who are willing to maintain on-farm crop diversity can be supported in their efforts in a number of ways, by national governments and local government as well as by individual development projects and programs.

Development policies

We saw earlier that the economic development process itself may promote on-farm crop diversity, although this point requires further research. Our results suggest that, where farmers' livelihoods are already buffered to a certain extent against outside pressures by the economic, sociocultural, and environmental resources at their disposal, encouraging families to maintain greater levels of crop diversity on-farm may well be possible. Examples of policy changes that might promote on-farm diversity include the provision of marketing facilities for all crops and crop varieties, so that farmers can sell a range of crops and varieties for cash, not only maize. This may involve upgrading the general level of transport and market infrastructure, and manipulating crop pricing and marketing policy, as well as input and credit policy. Other policy changes might include investing in the development and dissemination of processing equipment for different crops and varieties, so that non-maize crops do not have the disadvantage of having to be processed by hand, as at present in Zimbabwe. In particular, this might encourage more women to become interested in maintaining crop diversity on-farm: we saw earlier that the great amount of labor associated with the growing and processing of many non-maize grain crops has discouraged women from growing these crops, despite their interest in them for nutritional reasons.

Our results also suggest, however, that it is important for development policies not to focus exclusively on integrating all crops and crop varieties into the market economy, but to recognize the role played by crop diversity in providing household food security, as well as the role of different crops and crop varieties in local bartering and exchange at peak periods of food shortage.

We described the phenomenon of "maize-minded" farmers earlier in the chapter. Maize-mindedness arises from a combination of powerful economic forces as well as from changing cultural attitudes; therefore it may not be possible to reverse farmers' maize-mindedness on a wide scale. Nonetheless, it might be helpful for on-farm conservation, for the reasons outlined above, if these kinds of policy changes were made in order to allow different crops and crop varieties to fulfill a supportive role to maize in farmers' livelihood strategies.

Agricultural extension policy

We saw earlier how contact with extension services appears to have a positive effect on the number of crops and crop varieties grown by a household. This implies that increasing the number of households in contact with extension services would be beneficial for crop diversity. In the present era of pressure on government budgets, it may not be feasible to do this by increasing the number of government extension agents, but alternative approaches could be tried; examples include delivering extension services through preexisting community groups, identifying local farmers as "para-extensionists," or using mass media (Christoplous and Nitsch 1996).

Although our results suggest that contact with extension services is positively correlated with some aspects of on-farm crop diversity, it is important to remember that the traditional extension emphasis on promoting maize monoculture and pure-stand cultivation is still official policy in Zimbabwe. It might be helpful in encouraging farmers to maintain crop diversity on-farm if the extension service increased the extent to which it directly promotes on-farm crop diversity. Appropriate changes might include, for example, changing the criteria by which "Master Farmers" are judged by the extension services, away from maize monoculture and pure-stand cultivation toward production of numerous crops and crop varieties; also developing relevant extension messages for non-maize crops, which have often been relatively neglected to date; and adding competitions for on-farm crop diversity to the usual competitions organized by the extension services at local agricultural shows.

Plant breeding policies

We discovered during the participatory rural appraisal exercises that crop diversity is much more central to farmers' existence than has been previously acknowledged. Our results suggest that farmers attach great importance to having a wide range of crop varieties on-farm, to give them the flexibility not only to cope with an unreliable, resource-poor environment, but also to manage environmental variability to their best advantage.

Participatory plant breeding seeks to deliver planting material that is closely in line with farmers' needs, more quickly than is possible through conventional plant breeding. It can do this in a number of ways, including providing farmers with a relatively large amount of material, from which they can select according to their own requirements, discarding material which they consider to be unsuitable (Witcombe et al. 1996). This implies that participatory plant breeding can make a real contribution to supporting farmers in their efforts to maintain a wide range of crop varieties on-farm. So far, participatory plant breeding has not received as much attention in Africa as it has in Asia (Sperling and Loevinsohn 1996), but our results imply that it could usefully be encouraged in this region as well. Providing farmers with information on how to select for desired characteristics, in addition to providing the planting material itself, would give farmers even greater control of the breeding process, thereby further reducing their dependence on "ready-made" finished varieties released by formal sector plant breeders.

Seed supply policies

Our results suggest that seed security — both access to sufficient seed and access to seed of desired crop varieties — is an important variable encouraging farmers to maintain a large number of crops and crop varieties on-farm. Taking steps to support the availability of seed and varieties locally is, therefore, likely to be useful in helping farmers to maintain on-farm crop diversity. Our participatory rural appraisal exercises on mobility-mapping show that effective exchange of seed at the local level depends on different sections of the community interacting with each other, so steps to facilitate seed security could include encouraging increased contact between different sections of the community. In particular, we saw earlier that richer farmers as well as poorer farmers, and also older farmers and those with a position of authority within the local community, all tend to be willing to maintain crop diversity on their farms, albeit in different ways. This implies that particular efforts should be made to encourage these groups to participate in local level seed exchange. Some examples of how contact between different sections of the community could be facilitated are given in the section on indigenous culture below.

Another way of strengthening local level seed exchange could be to support local seed banking, including investment in local level seed bulking, processing, packaging, and distribution facilities. In the course of our research, we found that a pilot project to build a small number of community seed banks in Mutoko has demonstrated that providing a safe place to store seed within the community can significantly increase the availability of desired varieties locally. We also found that local seed banks can have a demonstration effect, encouraging more farmers to experiment with maintaining crop diversity on-farm, by ensuring that those who are interested have a ready source of seed and information about different varieties.

Our results suggest that supporting links to non-local sources of diversity could further strengthen on-farm crop diversity. Such sources include the formal seed sector, producing modern varieties, and national or regional gene banks holding indigenous landrace material. One example of the former from Zimbabwe is the modern sorghum variety SV2: while popular in its own right, our results show that it is also grown in Mutoko and Mudzi in mixed stands with local sorghum varieties specifically to permit introgression with these local varieties.

Although our results suggest that having enough seed during droughts is positively correlated with maintaining on-farm crop diversity, evidence from elsewhere (see, for example, ODI Seeds & Biodiversity Programme 1996) suggests that handouts of seed of inappropriate varieties and other inputs after drought or armed conflict can have a very negative effect on on-farm crop diversity. This implies that particular care needs to be taken when designing and implementing emergency seed distributions, to ensure that the seed supplied is appropriate to the local farming system.

Indigenous culture

We suggested above that one important way in which local seed exchange can be strengthened is by encouraging interaction between those in the community who have on-farm crop diversity and those who need it. We suggest that supporting indigenous culture, in terms of both community organization and cultural attitudes, is an important means of doing this. For example, our participatory rural appraisal exercises revealed that in Mutoko and Mudzi it is often older women who are interested in keeping seed of different crop varieties, and who retain the knowledge of how to plant and care for them. In the past, these older women have been able to obtain this seed by bartering for it with handicrafts such as clay pots, which are needed by other members of the community. However, as the local economy has become more cash-based, the demand for these handicrafts has declined, and therefore so has the means of obtaining seed of different varieties. In this context, a useful activity could be to find ways of encouraging the local barter economy.

Our results also suggest that it is also important to support traditional cultural attitudes toward crop diversity, because these usually place a high value on diversity, but may be under threat in "modern" society. This support could be provided by, for example, community meetings, discussions, and plays which present positive images of traditional "cultural identity," and of local indigenous knowledge, such as women's knowledge about how to care for seeds. Other strategies could include encouraging older people to pass on their knowledge of the value of on-farm crop diversity to younger family members; supporting the role of local traditional authorities; and providing opportunities for "study groups" where members can meet and discuss issues around the topic of on-farm crop diversity.

Conclusions

Managing vs. conserving on-farm crop diversity

Notwithstanding the discussion in the previous section, which suggests that pro-diversity farmers can be identified and supported, we suggest that our results add to the mounting international evidence (see, for example, Berg 1996) that farmers *manage* rather than *conserve* on-farm crop diversity. In other words, farmers do not preserve a static portfolio of crops and crop varieties on their farms, nor do they prevent introgression from neighbors' fields, field margins, fallow fields, or areas where wild crop relatives grow, but rather they import and discard diversity in a dynamic fashion, according to their needs in any given period of time.

We suggest that most "on-farm conservation" projects that succeed in motivating farmers to preserve individual crops and crop varieties are reliant on compensating farmers, usually through payment.[8] This can result in reducing the dynamism and flexibility in the farming system, because farmers do not usually view on-farm crop diversity in a static way, but rather as a dynamic part of their farming system that can be manipulated as part of their constant struggle to achieve sustainable livelihoods. Consequently, we suggest that it is not possible to achieve long-term conservation of individual crops and crop varieties on-farm using farmers' existing management strategies, although it may be possible to support farmers in maintaining a dynamic portfolio of on-farm crop diversity.

Different actors, different objectives

During the course of this research, we have observed that a wide array of participants are involved in the conservation of crop diversity on-farm, including gene center scientists, non-governmental organization development workers, and farmers themselves, and each of these groups has a different understanding of how farmers can contribute to crop diversity conservation. We suggest that there should be more debate between these different groups, in order to determine what can be realistically expected from on-farm conservation of crop diversity.

As described earlier, farmers are looking to *manage* on-farm crop diversity with the aim of optimizing their overall livelihoods. This management may involve broadening or narrowing the range of crops and crop varieties used on-farm, and the balance between landrace material and modern varieties, according to the economic, sociocultural, and environmental circumstances of the individual farm family.

Gene center scientists and plant breeders, however, may look to *preserve* particular crops and crop varieties on-farm, as a means of ensuring that the maximum possible range of plant genetic resources is available today and in the future (Maxted et al. in press; Hodgkin et al. 1993). In this context, they are likely to be working to ensure that no genetic material leaves the farm, either discarded by farmers or through natural processes. In addition, they are often interested in ensuring that there is no introgression, which might "contaminate" the genetic material already on-farm. On the other hand, non-governmental organization development workers may look to *maintain* on-farm crop diversity, primarily with the objective of making farmers' livelihoods more sustainable. This was a point made several times by non-governmental organization representatives at the workshop on *Supporting Diversity Through Sustainable Livelihoods: What Are Farmers' Choice?* that we held in Harare in November 1996.

However, only a minority of non-governmental organization development workers are *directly* involved in activities to maintain on-farm crop diversity. Many more are involved in activities which affect on-farm crop diversity by *indirect* means, such as drought relief, agricultural technology transfer, or local level seed supply, without realizing that these activities may have a significant effect on on-farm crop diversity. We suggest that our results call into question the continued widespread promotion of "technology packages" of modern varieties and agrochemicals by many development projects. While increased food production through the use of these packages is proposed as a solution toward improved food security for growing populations, use of such packages results *de facto* in increased penetration of formal sector science into rural peoples' knowledge systems. The very processes by which on-farm crop diversity is managed may be undermined, and may become static or redundant given the pressures smallholder farmers face. We suggest, therefore, for conservation policies to be effective in maintaining on-farm crop diversity, farmers' knowledge as derived from resource management practices should be seen as management *of* shifting boundaries created by economic, sociocultural, and environmental variables, rather than as management *within* boundaries set by a technology package.

Definitions of crop diversity

Gene center scientists and plant breeders are most concerned with diversity at the *molecular* level (for example, whether particular alleles are absent or present in a population) (Hawkes 1991; Dempsey 1996).[9] Farmers, however, are most interested in *morphological* and *agronomic* variation, and how this can be used within the farming system to achieve sustainable livelihoods. Farmers can only easily recognize variation that can be seen by the human eye. Non-governmental organization development workers may not have been trained specifically in crop biology, and so may not have any under-standing of the implications of the differences between "seeds" and "vari-eties," between "landraces" and "modern varieties," and between crop diversification and varietal diversity.

Our research has shown us that there is wide variation between the different actors involved, in their respective interpretations of what "crop diversity" means in practical terms. Again, we suggest that there should be more debate between these different groups, so that each understands what the others are expecting from on-farm conservation of crop diversity.

Notes

1. Using funding provided by SIDA, Sweden and Darwin Initiative of UK De-partment of the Environment. We gratefully acknowledge the support pro-vided by these two institutions, but responsibility for the final analysis rests with the authors alone and does not necessarily reflect the views of SIDA, the Darwin Initiative, ODI, or DR&SS.
2. This work is described in fuller detail in van Oosterhout and Cromwell (in press) and van Oosterhout (in press). Readers are strongly advised to refer to these sources if further explanation of the methodology used and results obtained are required.
3. Copies of the proceedings of this workshop are available on request from ODI.
4. The information in this section is taken from surveys conducted by the Sor-ghum Landrace Study and the Darwin Initiative for In Situ Conservation in Zimbabwe, and from ENDA (1995) and GDI/ENDA (1994).
5. Most of both districts lies within Zimbabwe's semi-arid Natural Regions IV and V.
6. Farmers who follow a set of cultivation practices recommended by the gov-ernment agricultural extension department.
7. Although this assumption is widespread, we have not found any published evidence that proves it.
8. Mostly through direct personal observation of projects and discussions with project staff, but also through a global survey of on-farm conservation activ-ities summarized in Cooper and Cromwell (1994).
9. We are grateful to Louise Sperling who was the first person who encouraged us to think of actors' differing definitions of diversity in the terms which follow.

References

Berg, T. 1996. *Dynamic Management of Plant Genetic Resources: Potentials of Emerging Grass-roots Movements*. Study No. 1. Rome: FAO Plant Production and Protection Division.

Christoplous, I. and U. Nitsch. 1996. *Pluralism and the Extension Agent: Changing Concepts and Approaches in Rural Extension*. Department for Natural Resources and the Environment Publications on Agriculture, No. 1. Sweden: Swedish University of Agricultural Sciences.

Cooper, D. with E. Cromwell. 1994. *In Situ Conservation of Crop Genetic Resources in Developing Countries: The Influence of Economic, Policy and Institutional Factors*, ODI Discussion Paper (draft). London: ODI.

Dempsey, G. 1996. *In Situ Conservation of Crop and Their Relatives: A Review of Current Status and Prospects for Wheat and Maize*. CIMMYT Natural Resources Group Paper 96-08. Mexico: CIMMYT.

ENDA. 1995. *Adaptive Strategies for Sustainable Livelihoods in Arid and Semi-arid Lands*. Zimbabwe: ENDA, for International Institute for Sustainable Development (IISD), Canada.

GDI/ENDA. 1994. *Facilitating Sustainable Development in Zimbabwe: Key Factors and Necessary Incentives*. Berlin: German Development Institute; Zimbabwe: ENDA.

Guveya, E. 1996. Crop diversity, crop diversification and household food security. Paper presented at the workshop *Supporting Diversity Through Sustainable Livelihoods: What are farmers choices?* 5–7 November, Harare, Zimbabwe.

Hawkes, J.G. 1991. *Genetic Conservation of World Crop Plants*. London: Academic Press.

Hodgkin, T., V. Ramanatha Rao, and K. Riley. 1993. Current issues in conserving crop landraces *in situ*. Paper presented at a workshop on on-farm conservation, 6–8 December 1993, Bogor, Indonesia.

Maxted, N., B.V. Ford-Lloyd, and J.G. Hawkes (eds.). 1997. *Plant Genetic Conservation: The* in situ *Approach*. London: Chapman & Hall.

Maxted, N., L. Guarino, B. Landon Myer, E.A. Chiwona, R. Crust, and A. Eastwood. In press. Towards a model for on-farm *in situ* plant genetic conservation.

MLARR. 1990. *Farm Management Survey of Zimbabwe*. Harare: Ministry of Lands, Agriculture and Rural Resettlement.

ODI Seeds & Biodiversity Programme. 1996. Seed provision during and after emergencies, *Good Practice Review* 4. London: Overseas Development Institute.

Page, S.L.J. and A. Chonyera. 1994. The promotion of maize fertilizer packages: a cause of household food insecurity and peasant impoverishment in high rainfall areas of Zimbabwe, *Development Southern Africa* 11:301–320.

Prain, G. and M. Piniero. 1994. Community curatorship of plant genetic resources in Southern Philippines: preliminary findings. In *Local Knowledge, Global Sciences and Plant Genetic Resources: Towards a Partnership*. Proceedings of an international workshop on user participation in plant genetic resources research and development. 4–8 May 1992, Los Banos, Philippines. Los Banos: UPWARD.

Sperling, L. and M. Loevinsohn. 1993. The dynamics of adoption: distribution and mortality of bean varieties among small farmers in Rwanda, *Agricultural Systems* 41(4):441–454.

Sperling, L. and M. Loevinsohn. 1996. *Using Diversity: Enhancing and Maintaining Genetic Resources On-Farm*. Proceedings of a workshop, 19–21 June 1995, New Delhi, India. Canada: IDRC.

UNEP. 1994. *Convention on Biological Diversity: Text and Annexes* UNEP/CBD/94/1 The Interim Secretariat for the Convention on Biological Diversity, Geneva.

van Oosterhout, S. In press. *On-Farm Conservation and Sustainable Livelihoods: Training and Methods Manual for Field Investigations.*

van Oosterhout, S. and E. Cromwell. In press. *On-Farm Conservation and Sustainable Livelihoods: Lessons from Zimbabwe.*

Witcombe, J.R., A. Joshi, K.D. Joshi, and B.R. Sthapit. 1996. Farmer participatory crop improvement. I. Varietal selection and breeding methods and their impact on biodiversity, *Experimental Agriculture* 32:445–460.

chapter ten

In situ *conservation and* intellectual property rights

Carlos M. Correa

Introduction

The conservation and use of plant genetic resources is described as a "system" wherein different agents play distinct roles. The creation of knowledge by indigenous/traditional farmer communities is characterized and compared to knowledge production in the "science" and "technology" systems. Intellectual property rights are currently applicable to downstream activities, while knowledge generated upstream is deemed to be in the public domain, despite its economic value. Numerous approaches aim to extend or develop alternatives to intellectual property rights. It is argued that many proposals are grounded in a conception of "natural rights," which provides an inadequate justification for a positive regulation of plant genetic resources and indigenous knowledge. An instrumental conception of intellectual property rights is needed to clarify the objectives that society pursues through protection (or other policies) and ensure that the established legal mechanisms are adequate to attain the intended goals. For this purpose, a number of clarifications and distinctions are necessary.

This chapter considers the concept of Farmers' Rights, still undefined with regard to its scope and content. It also addresses the difficulties inherent in developing special intellectual property rights for the protection of traditional farmers' varieties (landraces), as an extension of plant breeders' rights. Finally, an alternative legal approach is proposed based on a *sui generis* regime — inspired by the protection of trade secrets — that may be developed at the international level.

The plant genetic resources system

Conservation (*in situ* and *ex situ*), research and development, and utilization of plant genetic resources are components of a complex system in dynamic interaction. Such an interaction is based on market and non-market relationships among different types of agents with specific functions within a system called the "Plant Genetic Resources System" (Figure 10.1). Agents in the plant genetic resources system include traditional farmers and indigenous communities, collectors and curators (conservation subsystem), research institutions (research and development subsystem), breeders and seed companies (commercial breeding/production subsystem), and farmers (agricultural use subsystem). Each of these groups performs different functions within a particular framework of customary and legal rules.

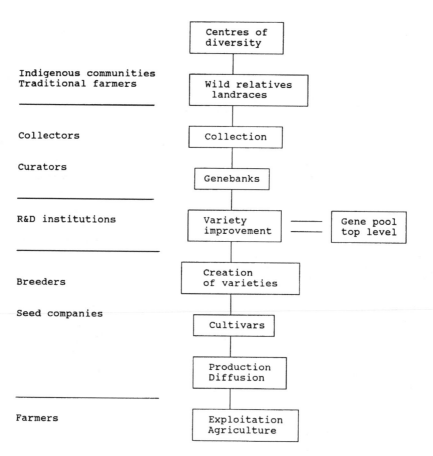

Figure 10.1   The PGR system.

Among the different agents who are directly involved with plant genetic resources are indigenous peoples, collectors, research institutions, breeders, seed companies, and farmers. Indigenous peoples and traditional farmers both conserve and use plant genetic resources system. The value of plant genetic resources is preserved and enhanced by their utilization for planting, seed production, and continuous selection of the best adapted local varieties (landraces). They generally interact among themselves on the basis of barter or exchange across the fence.

Collectors and curators collect and/or conserve and manage plant genetic resources, specifically with regard to their characterization, cataloging, evaluation, and pre-breeding. They interact with traditional farmers, research institutions, breeders, and seed companies. In most cases, such an interaction is based on non-market transactions. Traditional farmers are not paid a price for the value they deliver; breeders and seed companies are not charged a price for the samples they obtain. Research institutions utilize plant genetic resources to undertake basic and applied research, including agrobiotechnology, and to enhance existing varieties and the availability of genepools. Interaction with other agents in the system (traditional farmers, curators, breeders) is generally on a non-market basis. However, a strong trend toward protection of research results and increased linkages with private companies is the introduction of market-based transactions.

Breeders utilize plant genetic resources in breeding programs. They obtain materials and scientific information from traditional farmers and research institutions, generally on a non-market basis, and produce new or improved varieties for sale in the market. Intellectual property rights, wherever available, strengthen their market position and their ability to recover development expenditures. Seed companies utilize breeding results to propagate and sell seeds. They operate entirely within the market. Plant genetic resources are one of the intangible inputs in seed production, although these resources are not attributed a particular value, except where protected by intellectual property rights.

Finally, farmers who utilize improved varieties are at the end of the research/production chain. They benefit from the work realized, remunerated or not, within other subsytems. Their relationship with seed suppliers is market-based. Farmers both use and produce seeds, which they can reuse freely or in the framework of the "farmer's privilege,"[1] where applicable.

The indigenous/traditional knowledge system

The information and materials generated and used upstream in the *in situ* conservation subsystem are presently considered free goods. They belong to the public domain in that they are available to anyone without the permission of the developer/conserver and without any remuneration. The fact that indigenous/traditional knowledge belongs to the public domain does not mean that it is developed without intellectual effort, or that it is deprived of any value. Such knowledge is the result of a structured system of understanding, and certainly has an economic value although not necessarily a commercial value in a market.

The knowledge of indigenous and traditional farmer communities encompasses a set of different components (Box 10.1) that are part of a "traditional" knowledge system with its own epistemological foundations and practitioners (Shiva 1996:21). It is "an "organized, dynamic system of investigation and discovery that is of critical value to the sustainable maintenance of earth's diversity" (Shiva 1996:13). A main feature of this system is that knowledge is produced collectively. Innovation is "accretional" and "informal" and takes place over time. "The knowledge evolves as it modifies, adapts and builds upon the existing 'knowledge'" (Shiva 1996:23).

Box 10.1   Components of Traditional/Indigenous Knowledge

(a) Technologies and know-how relevant to the identification characterization and monitoring of ecosystems, species, and genetic resources:
   (i)   traditional knowledge about local ecosystems
   (ii)   traditional knowledge about ecosystem function
   (iii)   traditional knowledge of territories and habitats
   (iv)   traditional and advanced taxonomies
   (v)   uses, both traditional and current
   (vi)   traditional knowledge of technologies to determine species and genetic resource status and of population norms over time
   (vii)   traditional techniques for communication and information transmittal
(b) Technologies appropriate for the *in situ* conservation of components of
   (i)   traditional knowledge and technologies for *in situ* conservation
(c) Technologies for sustainable use of biological diversity and its components:
   (i)   spiritual and cultural uses
   (ii)   traditional medicine production techniques
   (iii)   natural resource management with the use of indigenous knowledge and technologies
   (iv)   methodologies for evaluation of biological divesity, including non-economic values such as existence, religious, ethical, and cultural values.

Source: UNEP/CBD/COP//3/19, 1996, p. 11.

There are many clear differences between the "traditional" knowledge system and the "scientific" and "technological" systems as they are known today (see Gibbons et al. 1994). Such differences relate to factors such as the process of creation of knowledge, the kind of creators, the methods used, the systems of compensation and validation, the level of codification (formalization) of knowledge, the existence of property rights, and the modes of diffusion. As illustrated in Table 10.1, there are, however, also some similarities. The term Science and Technology (S&T) is used forthwith to refer to these scientific and technological systems.

Table 10.1   Knowledge Creation in the Traditional, Science, and Technology Systems

Knowledge System	Developers/ Creators	Methods	Reward System	Validation/ Evaluation	Calification	Diffusion
Traditional	Commu- nities	Empirical	—	Use	—	Free
Science	Individuals/ groups of researchers	Scientific	Reputation for first discovery	Evaluation by peers	Codified (public- ations)	Free
Technology	Individuals/ employees	Empirical/ scientific	Appropri- ation of rents	Market success	Codified tacit	Restricted subject to prior authori- zation

Knowledge creation and compensation in the traditional, science, and technology systems

A comparison of the "traditional" system with the two other systems indicates clear differences with regard to who creates knowledge and the methods of validation, compensation, and appropriation. Traditional knowledge, as mentioned above, is created by communities, while science and technology are developed by individuals, teams of researchers, or employers hired by firms. Traditional knowledge is validated by the use of knowledge within communities, while scientific knowledge is validated by peer evaluation, and technology by its use and success in the market. There is no formal reward mechanism in the traditional system, whereas reputation given by first discovery is the dominant means of reward in science, and appropriation of rents in the technology system.

In other aspects, however, some similarities between traditional and S&T knowledge creation emerge. For instance, the creation of technology in both is essentially cumulative in nature. Technology advances both by means of "radical" innovations as well as minor, "incremental" innovations that play a key role in technological change, including, for example, dynamic areas such as electronics. Technology sources include both scientific inputs as well as empirical inputs that generally result from "learning by doing." Important components of technology often are not codified but maintained in a "tacit," informal form (Cassiolato 1994:279). Finally, in both the traditional system of knowledge and in modern science, public, nonproprietary knowledge is created. In neither case is created knowledge appropriated under intellectual property rights; diffusion of knowledge is free and without restrictions. In other words, science in S&T and traditional knowledge are in the realm of public domain. How much of the public domain will survive in an era of expanding intellectual property protection is an important issue to be further investigated.

Economic value

The economic value of indigenous/traditional knowledge is receiving growing recognition. Knowledge about medicinal uses of plants, for instance, saves substantial research costs to pharmaceutical companies and provides insight into unique therapeutic options (Shiva 1995:130). Indigenous/traditional knowledge has economic value, although the economic value of biological diversity for agriculture is difficult to assess (Brush 1994). A comprehensive theoretical framework and solid empirical evidence are missing. Some fragmented evidence is available on the benefits obtained by recipients of plant genetic resources. For instance, a detailed study on the value of rice landraces to Indian agriculture showed that they contributed 5.6% to India's rice yields, with an estimated value of $75 million (National Research Council 1993).

The economic value of plant genetic resources may be analyzed, in marginal terms, on the basis of the opportunity cost of the conversion of biodiversity to specialized production. The rationale for this approach is that, while conserving landraces, traditional farmers are deprived of obtaining higher productivity and income associated with the use of modern varieties. There is, hence, a global value determined by the differential in the average yield between the use of land in a traditional versus a specialized form of production (Swanson et al. 1994:25). The value of genetic diversity, however, is not limited to the opportunity cost borne by traditional farmers. Maintenance of biological diversity in farming systems generates value for the global community which is determined by three additional components:

1. "portfolio effect," the static value of retaining a wide range of varieties and methods of production, which reduces the risk of variable production;
2. "quasi option value," based on the value of the future flow of expected information to be generated by the retained diversity;
3. "exploration value," or the value of retaining the evolutionary processes of varieties and the opportunity to discover new traits and characteristics (Swanson et al. 1994:26).

The availability of germplasm enables farmers to face changes in the environment, or occurrences of disease or pests ("quasi option" value). The "exploration value" may be of particular importance for biotechnology-based industries, which can exploit genes of particular agronomic interest. Consumers, finally, benefit from a reduced risk of variability in production ("portfolio value") and from improved and increased production.

Farmers holding landraces, thus, create an economic value. They are currently, however, unable to appropriate it for the purposes of generating revenue as income. In economic terms, they generate externalities as providers of a "public good." This does not mean that other agents could not benefit and eventually appropriate downstream the values so created. The direct beneficiaries of the value created by the nonconversion of land from traditional to specialized uses are those able to utilize the germplasm so conserved, including farmers and breeders of all countries — not only of the country where the relevant landrace is located. There is a strong inter-dependence among countries with respect to plant genetic resources, and in most cases these resources are found in several countries; distribution is not constrained by national boundaries. From an economic point of view, plant genetic resources have a "global" value, the realization of which benefits farmers, breeders, and consumers all over the world. Thus, plant breeders and seed companies may capture the rents of plant genetic resources, which they have incorporated into new or improved varieties that become pro-tected by intellectual property rights.

Intellectual property rights downstream

Only a minor portion of the materials maintained *in situ* or in *ex situ* collec-tions enter the research and development subsystem. If this is the case, and depending on the characteristics of the research and development results, plant genetic resources that have been distributed from *ex situ* collections may — but not necessarily — give rise to claims of intellectual property rights. Thus, public institutions have generally produced improved varieties, including hybrids, which were released for free use by farmers. Recent trends toward "privatization" of agricultural research and the need to secure funds for sustaining research and development projects have led, however, to a growing use of the intellectual property rights system by public institutions.

Intellectual property rights play an important role in the commercial breeding/production subsystem. The availability of intellectual property rights may stimulate the development of "modern," "commercial" varieties which comply with the requirements (particularly stability and uniformity) imposed by breeders' rights regimes. In theory, the availability of intellectual property rights stimulates research to benefit the public by the release of improved crop varieties. In this way, private crop breeding can both com-plement and compete with public crop breeding in national and international research.

Intellectual property confers, in general, exclusive rights with respect to the use of information in different areas of knowledge. Some types of intel-lectual property rights are particularly relevant to agriculture. These rights include breeders' rights, patents, utility models, trade secrets, and geograph-ical indications. Each of these types of intellectual property rights applies to different subject matter as described in Table 10.2.

Principal intellectual property rights applicable to agriculture

Breeders' rights are a type of intellectual property related to propagating materials of plant varieties. They constitute the single specific, *sui generis*

Table 10.2   Main Intellectual Property Rights as Applicable to Agriculture

Title	Subject Matter
Breeders' rights	Propagating materials of plant varieties
Patents	Mechanical, chemical, and biological inventions
Utility models	Improvements in machinery and tools
Trade secrets	Undisclosed information of commercial and technical value (e.g., hybrids)
Geographical indicators	Name of country, region, or locality where agricultural products originate

protection available in the field of agriculture. Breeders' rights have been adopted by most developed countries but by only a few developing countries.[2] The UPOV (Union for the Protection of Plant Varieties) Convention provides an international framework for the protection of said varieties. The World Trade Organization (WTO) rules and, in particular, the Agreement on Trade Related Aspects of Intellectual Property Rights (TRIPs Agreement) will, however, lead to the creation of intellectual property rights regimes, including breeders' rights, in all countries that join these agreements. Few countries will remain outside of WTO and the TRIPs Agreement. The impact of the introduction of breeders' rights on seed production and research and development is still relatively unexplored (see Jaff and van Wijk 1995).

Patents are conferred in many countries to protect inventions relating to plants and animals (including genetic materials). There are, however, important differences among national laws on the subject matter of protection. The TRIPs Agreement allows member countries to exclude plants and animals from patentability, but plant varieties need to be protected either by patents, by an "effective *sui generis*" regime or by a combination of both. Developing countries have a transitional period of 5 years to comply with the standards of the TRIPs Agreement.

The application of patents to plant parts, including cells and genes, has been accepted in many countries. This remains, however, a controversial issue, particularly with regard to the patentability of materials that pre-exist in nature and are just isolated and purified, or slightly altered, in order to be claimed as an "invention." Patents may also apply to many other products (e.g., agrochemicals, equipment) used in agriculture, including biotechnology-based products such as vaccines and biopesticides (Wegner 1994). Patents, in sum, may have a wide impact on various aspects of agricultural activities.

Utility models are relevant for the protection of functional improvements in agricultural machinery and other tools. The requirements to be met in order to obtain these titles, in terms of novelty and inventiveness, are normally lower than those for patents: utility models apply to "minor" innovations. Unlike patents, such models are generally applied for and conferred to nationals of the countries of registration, rather than to foreigners.

Trade secrets protect undisclosed information of commercial and technical nature, as long as it remains secret and the possessor has taken reasonable measures to prevent its disclosure. As in the case of patents, trade secrets apply to many products relevant to agriculture, including the processes involved in the production of many biotechnology-based products. One of the main fields of application of trade secrets in agriculture relates to hybrids, such as maize. In this case, "technical protection" (Jullien 1989) is high, in the sense that pertinent information cannot be easily obtained from the product itself (unlike the case of self-pollinating varieties). Though a trend toward the protection of parental lines via breeders' rights can be observed, trade secrets are still the principal means of protection for such types of seeds.

Finally, it is necessary to mention geographical indications among the intellectual property rights relevant to agriculture. A geographical indication consists of the name of a locality, region, or country, which is used by producers located therein to indicate the geographical origin of certain products. Such a use is subject to different requirements under existing domestic legislation. In order to be protectable, the characteristics or reputation of the products needs to be essentially attributable to a given origin (Bérard and Marchenay 1996).

Extending intellectual property rights to indigenous/traditional knowledge

As described above, despite its economic value, indigenous/traditional knowledge belongs today to the public domain. Intellectual property rights are only relevant in downstream activities, even if they benefit from values created upstream. The question to be addressed is whether intellectual property rights should be extended upstream to such a knowledge and, in that case, for what purpose, for whose benefit, and under which conditions. Thus far, the analyses and discussion are not generally clear with respect to the foundations and objectives sought with an eventual extension of intellectual property rights. As mentioned above, indigenous/traditional knowledge is composed of a number of different elements (Box 10.1), the application and value of which vary significantly. This frequently creates confusion about the specific knowledge component for which the creation of such rights is advocated.

There are both ardent proponents and critics of extending intellectual property rights to the knowledge of indigenous and traditional communities, including landraces. Those who are reluctant to create or who oppose the idea of creating a new form of intellectual property rights offer arguments based on both principles and practical reasons. Some indicate, for instance, that bringing communities and their resources into the fold of the market economy could overwhelm and ultimately destroy those societies (Nijar 1996a:24). This might be overcome by a "rights regime which reflects the culture and value-system of these communities" (Nijar 1996a:24).

It has also been argued that, given the difficulties inherent in establishing intellectual property rights protection for indigenous/traditional knowledge, legislation and international conventions should ensure that such knowledge, biological materials, and their derivatives are not subjected to any kind of property rights (Montecinos 1996:22). They should remain a part of "public domain" everywhere. This would imply that such knowledge and materials should be declared unpatentable in all countries, including those that currently permit the protection of different forms of biological inventions. Other analysts question the instrumental value of intellectual property rights in the field of indigenous/traditional knowledge. Brush, for instance, has noted that,

> [T]he sheer volume of different farmer varieties, the fact that genetic diversity crosses national boundaries, and the large amount of genetic resources already collected and placed into the international public domain pose serious difficulties for any one farm group or nation seeking to claim novelty or distinctiveness.... Landraces are likely to have very little commercial value because of breeders' strong preferences for well known genetic material rather than exotic and unknown material.
>
> The relative abundance of germplasm in public institutions also lessens the possibility that breeders will purchase crop germplasm from farmers.... The abundance of collected germplasm thus undermines a market based on intellectual property for crop genetic resources. There seems to be little chance that users will pay for unknown germplasm when they can obtain it without cost from international and open collections (Brush 1994:25-26).

Different alternatives to intellectual property rights have been proposed to deal with indigenous/traditional knowledge or some components thereof. This is the case, for instance, of proposals relating to "tribal" or "communal" rights (Greaves 1996), "community intellectual rights" (Gebre Egzibher, 1996a:38), "traditional resource rights" (Posey and Dutfield 1996), and, most notably, to Farmers' Rights as a means of compensating traditional farmers for their contributions to the *in situ* conservation of plant genetic resources.

The recognition of Farmers' Rights is one alternative, introduced by the FAO International Undertaking on Plant Genetic Resources, to compensate traditional farmers. FAO Resolution 5/89 defines Farmers' Rights as:

> [R]ights arising from the past, present and future contribution of farmers in conserving, improving and making available Plant Genetic Resources, particularly

those in the centres of origin/diversity. These rights are vested in the International Community, as trustees for present and future generations of farmers, for the purpose of ensuring full benefits of farmers and supporting the continuation of their contributions.

One of the objectives of Farmers' Rights, in accordance with the same Resolution, is to "allow farmers, their communities, and countries in all regions, to participate fully in the benefits derived, at present and in the future, from the improved use of Plant Genetic Resources, through plant breeding and other scientific methods." In developing this concept, the FAO Commission on Plant Genetic Resources has agreed that a number of questions remain open and need to be addressed. These include:

- the nature of the funding (voluntary or mandatory);
- the question of linkage between the financial responsibilities and the benefits derived from the use of plant genetic resources;
- the question of who should bear financial responsibilities (countries, users, or consumers);
- how the relative needs and entitlements of beneficiaries, especially developing countries, would be estimated; and
- how financial and local communities would benefit from the funding.

The concept of "Farmers' Rights" has received wide acceptance, as indicated by the results of the Leipzig Conference on the Conservation and Sustainable Utilization of Plant Genetic Resources for Food and Agriculture (17–23 June 1996), where the concept was reaffirmed. After considerable debate a general compromise was reached on this issue (Gebre Egzibher 1996a:6) but no firm and clear commitments have been made with regard to the form of implementation of such rights at the international, regional, and national levels.

The content and scope of Farmers' Rights has not yet been fully defined. Their eventual assimilation to intellectual property rights has been questioned, as the latter might undermine the free sharing of knowledge and resources among local communities and the world community. Furthermore, this is recognized as incompatible with the collective nature of innovation at the community level (Gebre Egzibher 1996a). Farmers' Rights are regarded as "some counterbalance to 'formal' intellectual property rights which compensate only for the latest innovation, without acknowledging that, in many cases, these innovations are only the last step in cumulative inventions carried out over many human generations, in different parts of the world" (Esquinas Alcazar 1996:4). For those who advocate the establishment of intellectual property rights, their argument recognizes intellectual property rights as an ethical imperative and/or a necessary tool to preserve biodiversity and prevent further erosion thereof. Within this line of thought, two trends may be identified. First, there are many proposals to extend the application of current modalities of intellectual property rights, or to amend existing laws and practices, in order to include certain components of indigenous/traditional knowledge. Such proposals include:

1. the application of geographical indications, copyright (protection of folklore) or other intellectual property rights (Correa 1994);
2. increasing the flexibility of the requirements for the protection of traditional plant varieties, by applying, for example, a broader concept of "uniformity" than that which is generally accepted under UPOV-like plant breeders' rights; and
3. introducing new requirements into existing laws, such as the obligation to declare in a patent application the origin of materials used to develop the invention.

Alternatively, there are proposals to develop options to existing intellectual property rights. Under this approach, different variants and new modalities of intellectual property rights, differing both in scope and possible forms of implementation, also exist. Their aim is the establishment of a general, comprehensive, *sui generis* regime on indigenous and traditional communities' knowledge, covering knowledge on, *inter alia*, medicinal plants, materials useful for agriculture, and cultivation practices.

In sum, several alternatives to deal with indigenous/traditional knowledge have been put forth. They range from explicitly and universally excluding the appropriation of biological materials and related knowledge, to the development of completely new rules. The following section discusses the grounds for new approaches and potential changes in the legal systems.

Grounds for new rules

The establishment of new rules or of a *sui generis* regime for the protection of knowledge held by indigenous and traditional farmers has often been grounded on the need to recognize pre-existing rights of indigenous/traditional communities (Tilahun and Edwards 1996). Such rights would seem to exist, under this approach, before and independently from positive law. Thus, the law would not create such rights, but only provide for the conditions under which such rights should be recognized and exercised. This type of approach has certainly brought attention to the issue, stimulated a wide discussion, and prompted many proposals. This approach, however, generally fails to define the purpose and rationale for the protection. "Natural rights" theory is incompatible with a positive, nonconfessional conception of law. It reassembles current claims by industrialized countries against copying and "piracy" as grounded on rights beyond political frontiers and economic systems (Oddi 1996:424). The "natural rights" theory has been widely criticized and dismissed both as a general justification of law (Kelsen

1991), as well as a specific ground for the granting of intellectual property rights (Penrose 1974; Oddi 1996:431).

With regard to intellectual property rights, for instance, summarizing Jefferson's views on patents, the U.S. Supreme Court in *Graham v. John Deere Co.* recalled that Jefferson:

> ... rejected a natural right theory in intellectual property rights and clearly recognized the social and economic rationale of the patent system. The patent monopoly was not designed to secure to the inventor(s) ... natural rights(s) in (their) discoveries. Rather it was a reward, an inducement, to bring forth new knowledge (Government of India 1995).

The "natural rights" theory, therefore, does not provide a solid justification as to why the society should establish protection. An alternative theory is needed to elucidate the intended objectives of such a protection and subsequent benefits for society. A number of objectives may be defined. One objective may be, for instance, to reduce or avoid conversion from traditional to commercial varieties. In this case, a system that compensates farmers for lost income may be required. Swanson argues that in the absence of a compensation mechanism, traditional/indigenous farmers would tend to substitute their own varieties with higher yielding commercial varieties (Swanson et al. 1994). Under this conception, property rights would not be intended to "reward" communities for their contribution, to the maintenance and development of landraces, or to create incentives for investments. In Swanson's view, protection would be mainly justified to avoid conversion to modern varieties. It is, hence, a "conservationist" theory, based on protection as a means to maintain the current levels of conservation.

If the objective were to motivate farmers to invest and innovate more than they currently do (for instance, in crops of particular importance), any potential regime should not limit itself to compensation for lost income. Where public goods are created, investments for producing them necessarily tend to be suboptimal, since their producers are unable to benefit from the rents such goods may generate. This is a typical market failure that justifies public intervention, as illustrated by the case of basic science (Nelson 1971). Another objective may be to obtain a fair share of the benefits generated by the use of communities' knowledge. This may generate new income and bring a neccesary element of justice, but the mechanism in and of itself would not ensure that the funds needed to make a certain level of investments are created.

In summary, there is no doubt about the justice of proposals aiming at some kind of compensation or protection for indigenous/traditional knowledge. However, it is necessary to clarify what society would intend to reach through protection, and how such goals can be realized. Once these aims are clarified, property or other rights may be devised as instruments to attain them. An instrumental approach to the issue means that the establishment of property or other rights should be considered as a means to effectively reach the proposed goals.

Intellectual property rights and conservation

Consideration of intellectual property rights and *in situ* conservation for agricultural purposes requires a number of additional conceptual distinctions. First, the protection of varieties maintained and improved by local/indigenous communities needs to be distinguished from the protection of unmodified genetic sequences as such. While the former relates to materials that have been improved over time, the latter refers to information which exists in nature and which constitutes a "natural capital" of countries where the respective resources reside. The establishment of a new category of "informational rights" to protect these has been proposed (Swanson 1995:169; Walden 1995:191).

Second, a distinction should be made between the protection of landraces, i.e., improved materials useful for agriculture, from the protection of traditional knowledge held by local/indigenous communities about the possible uses of certain plants, particularly for medicinal purposes. In the case of the former, the subject matter is well defined (even if lacking stability) and protection would be dependent on the physical and actual existence of a variety. In the latter, however, what is at stake is knowledge on the use of materials, and eventually on procedures to extract or apply them (such as in the case of the neem tree, the seeds of which have been used as a pesticide in India for hundreds of years). The problems posed by the protection of this type of knowledge are quite different from those relating to plant varieties as such.

Third, while considering an eventual form of intellectual property rights protection for landraces, the objectives of the protection sought should be clarified, as mentioned before. Intellectual property rights provide a tool for the appropriation of rents based on different kinds of intellectual efforts. If the objective of the protection were, for instance, to remunerate for past contributions made to mankind by traditional farmers, intellectual property rights will not necessarily be the appropriate tool. Even if the objective were to reward investments and facilitate the diffusion of innovations (more in line with the typical foundations of intellectual property rights), it is important to note that intellectual property rights are not the unique or necessarily the best (both privately and socially) means to achieve such an objective. Finally, the need to conserve plant diversity on-farm seems to be well accepted. There are, however, some major pending questions. The desirable amount and composition of diversity to be conserved is unclear. Should traditional farmers remain limited to traditional varieties that preserve and enrich genetic diversity, but which are normally inferior to commercial varieties in terms of productivity and income generation? The adoption of commercial varieties has certain negative effects on biodiversity, but some undeniably positive economic and social effects as well. The question is, therefore, how to develop an agricultural policy, including intellectual property, that does preserve the required amount and composition of plant biodiversity, and at the same time allows poor, traditional farmers to benefit from higher yielding varieties.

Intellectual property rights for traditional varieties?

Should a specific form of intellectual property rights be recognized for traditional varieties (see Correa 1994)? To determine the feasibility and potential content of intellectual property rights for traditional varieties, the principal legal issues are examined below.

Definition of subject matter

Although modern techniques such as molecular markers allow for a detailed description of the heritable material of plants and plant populations, it seems difficult (if not impossible) to define individual landraces, which continuously evolve. If adopted, a system of protection should be based on the material existence of certain germplasm (as in the case of breeders' rights).

Under the UPOV regime, a variety cannot be protected if it was commercialized for more than 1 year before protection is sought, in the country where the application is filed (article 6). Therefore, the applicability of a UPOV-like standard of novelty to landraces is problematic since the landraces in question may have been used by communities long before any protection is sought. The UPOV uniformity requirement also poses a great obstacle. By their nature, traditional varieties are continuously evolving; they lack the stability and uniformity characteristic of modern varieties. The uniformity requirement may be relaxed to some extent, as in the case of the Austrian law (1993) on plant breeders' rights. According to this law, a plant variety is homogeneous when "its individuals, as a whole or with respect to a given distribution, are sufficiently uniform in the expression of each relevant characteristic, notwithstanding a small number of variations." Landraces typically present a high degree of diversity, which prevents a proper identification of the eventually protectable subject matter.

Territorial validity of rights

Patents and breeders' rights are territorial rights, in the sense that they only are valid in the countries where registration has been obtained. The main problem in this respect is the occurrence of the same landraces in several countries. To whom should the rights be accorded? Identifying the title holder is likely to be one of the main problems of developing intellectual property rights protection for landraces. Landraces generally have no single origin and they result from the interaction of multiple landraces over time.

A second problem created by territoriality is that in order for a community holding a right to obtain protection outside its own country, it would have to obtain similar protection in the third country, if recognized. This poses an operative burden of how to secure protection abroad. The effectiveness of any system of protection would depend on its recognition at the international level, and not only in one or a few countries.

Operationalizing the system?

Issues such as examination and registration of landraces should be further analyzed. If a regime of protection for "landraces" based on the concept of breeders' rights were developed, it would require the establishment of administrative structures for examination and registration of protected materials, resulting in potentially high transactions costs for governments and users of the system.

Availability to potential beneficiaries

Another key issue is the extent to which a system of protection would actually operate in favor of its intended beneficiaries. If the requirements of novelty or uniformity under breeders' rights legislation were relaxed to allow for the protection of "landraces," greater benefits may accrue those who are well positioned technically and financially. For example, seeds and biotechnology industries are likely to gain more from a system based on breeders' rights than indigenous communities.

Enforceability

Availability of rights is useless if the system is not enforceable. Enforceability depends on the ease with which material can be copied and on the capacity to monitor the use and eventual infringement of rights. An additional problem is financing the potentially high costs of administrative and judicial procedures required to stop infringement and obtain compensation for damages.

There are a number of complex issues to be considered for the extension of breeders' rights protection to landraces. The eventual establishment of exclusive, monopolistic rights, as conferred under plant breeders' regimes may, moreover, be essentially incompatible with communities' cultures and practices. A possible alternative to this approach, based on a non-monopolistic means of protection, is described in the following section.

A sui generis *regime*

The adoption of a *sui generis* regime on indigenous/traditional knowledge is conceivably one of the steps that may be taken at the national level and internationally to deal with the issues described above. The review of the TRIPs Agreement by the year 2000 (as noted in article 71.1) may provide an opportunity for developing international minimum standards on the matter. One of the legal foundations for such a regime may be found in article 8 (j) of the Convention on Biological Diversity adopted in 1992. In accordance with said article, traditional knowledge must be promoted and made more widely available, but knowledge must be used by others only with the "approval and involvement" of the original holders of that knowledge and the communities concerned. Communities should receive a fair share of the benefits from the use of their knowledge (Government of India 1995).

The definitional constructs of the TRIPs Agreement dismiss the knowledge systems and innovations of indigenous communities and farmers (Nijar 1996b). Nothing in the TRIPs Agreement, however, prevents member countries from establishing other forms of intellectual property rights protection (or even to increase the standards of protection). What members cannot do is provide protection below the minimum standards set forth by the Agreement. Thus, members may provide for the protection of utility models, a modality of intellectual property rights that is suitable to the type of innovations that prevail in developing countries, which was ignored in the TRIPs Agreement. In fact, many developed countries (e.g., Spain, Germany, Japan) and a growing number of developing countries (e.g., Brazil, Mexico, Uruguay, Argentina) provide this type of protection. Similarly, nothing in the Agreement prevents member countries from expanding the concept of plant varieties that may be protected under breeders' rights, or from establishing new forms of protection for indigenous and traditional farmers' knowledge.

With regard to the extent to which such rights are established on a national basis, member countries that recognize such rights could not enforce them in other member countries that do not. The same occurs, in fact, with a patent obtained in country A that has not been registered elsewhere. The invention simply belongs to the public domain, except in country A. This is a result of "territoriality" principle as applied to intellectual property rights.

The "informality" (non-codified), cumulative and predominantly incremental, nature of innovation in the traditional knowledge system is not a unique feature of this system. These characteristics are also present in the research and development system. Though patent rights do not apply when the created knowledge is not novel and nonobvious, other means of intellectual property rights protection, notably trade secrets (or "undisclosed information") regimes allow for the protection of routine, non-novel, non-codified (tacit) unregistered knowledge. Further, secrecy does not need to be "absolute," and protection lasts indefinitely, until the knowledge loses its secret nature. Trade secrets protection may apply to knowledge with both actual or potential commercial value.

The paradigm of trade secrets protection, therefore, provides a model on which a *sui generis* regime for protection of some kinds of traditional knowledge may be based. In some cases, knowledge (e.g., shamans' knowledge of medicinal plants) may qualify for straight protection as "undisclosed information" if it has been kept secret and other conditions for protection are met. In most cases, however, the knowledge may have been diffused to other communities, and no measures may have been taken to protect it from disclosure. Whether this knowledge would have any commercial value or not is difficult to ascertain because commonness and market value of traits are inversely related.

Developing a *sui generis* regime, faces important problems, as stated in a declaration of the government of India at the World Trade Organization Committee on Trade and Environment (Government of India 1995):

> New legislation and codes of conduct, including changes in the notion of "trade secrets" may be needed to ensure that the communities that are the source of this knowledge receive benefits from its exploitation. This is admittedly a difficult task since traditional communities do not usually have a legal identity and the knowledge concerned may not be confined to a single village or group, posing problems of deciding precisely who should derive the benefits and how (Government of India 1995).

A *sui generis* regime, if developed, should be applicable to all kinds of knowledge on biological materials held by communities, to the extent that such knowledge is not diffused outside said communities. It should cover knowledge on biological materials, including plant varieties, and on their production, use and conservation of this knowledge, which is possessed by indigenous or traditional farming communities (defined by national legislation).

Protection should not be based — as in the case of trade secrets — on an exclusive right (i.e., on an *ius prohibendi*). Protection should only grant the right to prevent knowledge of actual or potential commercial value, under the communities possession, from being acquired, used, or disclosed by others in a manner that is contrary to national rules on access or otherwise contrary to internationally accepted rules and practices of collection, transfer, and use of germplasm. The basic right should not be to prevent any third party (or another community) from the use of the protected knowledge if independently developed or otherwise legitimately obtained. Communities should, therefore, have the faculty to prevent knowledge of actual or potential commercial value, under the communities' possession, from being disclosed to, acquired or used by others without their prior informed consent in a manner contrary to internationally accepted practices of collection and transfer of germplasm.

"Internationally accepted practices" may be defined as those consistent with the Convention on Biological Diversity, the FAO Code of Conduct on the Collection and Transfer of Germplasm, and other international instruments developed in the future. In addition, to ensure that the rights of communities are not frustrated by the granting of patents or other titles on communities' knowledge, an effective *sui generis* should be complemented with a negative rule, according to which no intellectual property rights shall be conferred with respect to communities' knowledge, as described. In the case of infringement of this rule, the conferred title should be declared void, totally or partially, even in cases where the applicant did not know at the time of his/her application, that his/her claim was based on such knowledge. Knowingly or not, he/she was not the actual "inventor" and should not, therefore, benefit from a protection which rewards inventiveness and the contribution of new ideas to the pre-existing knowledge pool.

If a *sui generis* regime, as proposed, were established, national laws would be free to determine the means to ensure protection, including criminal and civil remedies, and how to empower communities to exercise their rights. The main features of a such a regime include:

- a definition of subject matter (knowledge, plant material), broad enough to cover any alteration, modification, or improvement, or a derivative which utilizes the knowledge of indigenous or traditional communities;
- a recognition of the informal, collective, and cumulative systems of innovation of indigenous peoples and communities;
- no requirement for novelty, inventiveness, or secrecy;
- no arbitrary time limit for protection;
- no registration, and therefore, no administrative structures;
- no obligation for communities' members to keep secrecy or change their traditional practices;
- a "non-monopolistic" provision of rights, which would permit the non-commercial use and exchange of germplasm within and among communities, and thus the legal possession and exchange of the same knowledge by different communities;
- freedom to determine, at the national level, remedies and sanctions in case of infringement.

The proposal outlined above does not solve all of the problems that arise in the attempt to extend protection to landraces and indigenous knowledge. Principal among the issues that remain are the determination of titleholder (who will exercise the rights?) and enforcement (how to ensure the respect of communities' rights and eventually stop infringement and obtain an economic compensation?). These are operative aspects that may be dealt with at the national level where there is a legitimate will to make progress on this subject. To ensure protection across national borders, the basic susbtantive rules should be adopted at the international level. The revision of the TRIPs Agreement in 1999 may provide such an opportunity.

Main conclusions

In situ conservation may be seen as a part of the world's plant genetic resource system. The knowledge and materials, including landraces, currently belong to the public domain. Indigenous/traditional knowledge is produced in accordance with patterns that present both differences and similarities with regard to the production of knowledge in the formal "science" and "technology" system. Though indigenous/traditional knowledge has economic value, it generally lacks a market or commercial value.

Intellectual property rights only apply to downstream activities, in different facets of agricultural activity. The extension of intellectual property rights "upstream" has both fervent proponents and detractors. If feasible at all, the application or development of a new title of intellectual property rights poses a fundamental question: should the basic conception of the intellectual property rights system (private appropriation of a public good) be extended to plant breeding of traditional farmers, or should their compensation be sought by other means, even if such means are based on market mechanisms?

The eventual development of new modalities of intellectual property rights for landraces presents numerous complex problems, particularly for the extension of plant breeders' rights to landraces. This implies the establishment of exclusive rights where free exchange has prevailed historically, in conflict with communities' cultures. If a form of protection for communities' knowledge, including landraces, is to be designed, a fundamental issue is defining the rationale and purpose of any future protection mechanism. A protection mechanism, whatever its nature and scope, is an instrument to attain certain socially valuable objectives and should appropriately balance the different interests at stake. Such objectives may include, *inter alia*, rewarding communities for past and present contributions; compensating farmers for non-conversion from traditional to modern varieties; and ensuring the sharing of benefits derived from the use of communities' knowledge. These objectives, if attained, may further the more general objective of enhancing *in situ* conservation of plant genetic resources.

The approach suggested in this chapter is based on the establishment of a "non-monopolostic" *sui generis* regime, inspired by trade secrets protection, whereby no registration would be necessary; all communities' knowledge of actual or potential commercial value may be protected against appropriation by non-legitimate means. Though many elements of such a regime would be determined at the national level, its recognition at the international would provide the necessary geographical coverage in order to ensure the effectiveness of the regime.

Notes

1. This is an exception generally allowed under breeders' rights regimes, which permits farmers to reuse, in their own exploitation, the seeds obtained from the utilization of protected varieties.
2. For example, in Latin America, Argentina, Chile, Mexico, Uruguay, and the Andean Group countries (Bolivia, Columbia, Ecudor, Peru, and Venezuela) currently recognize breeders' rights. Draft legislation is under consideration in several other countries, such as Brazil.

References

Bérard, L. and P. Marchenay. 1996. Tradition, regulation, and intellectual property: local agricultural products and foodstuffs in France. In *Valuing Local Knowledge: Indigenous People and Intellectual Property Rights*, S.B. Brush and D. Stabinsky (eds.). Washington, D.C.: Island Press.

Brush, S.B. 1994. *Providing Farmers' Rights through* in situ *Conservation of Crop Genetic Resources*. A report to the Commission on Plant Genetic Resources. University of California, Davis.

Cassiolato, J. 1994. Innovación y cambio tecnológico. In *Ciencia, Tecnologia y Desarrollo: Interrelaciones Teoricas y Metodologicas*, E. Martinez (ed.). Caracas: Nueva Sociedad.

Correa, C. 1994. *Sovereignty and property rights over plant genetic resources*. Report to the Commission on Plant Genetic Resources. Rome: FAO.

Esquinas Alcazar, J. 1996. The realisation of farmers' rights. In *Agrobiodiversity and Farmers' Rights*. Madras: Swaminathan Research Foundation.

Gebre Egzibher, T.B. 1996a. A case of community rights. In *The Movement for Collective Intellectual Rights*, S. Tilahun and S. Edwards (eds.). Addis Ababa: The Institute for Sustainable Foundation/The Gaia Foundation.

Gebre Egzibher, T.B. 1996b. The U.S. versus agrobiodiversity, *Third World Resurgence* 72/73:4–6.

Gibbons, M., C. Limoges, H. Nowotny, S. Schwartan, P. Scott, and M. Trow. 1994. *The New Production of Knowledge*. Stockholm: Sage Publications.

Glachant M. and F. Leveque. 1993. *L'enjeu des resources genetiques vegetales*. Paris: Les Editions de l'Environnement.

Government of India. 1995. *Statement by India on TRIPs, Environment and Sustainable Development at the WTO Committee on Trade and Environment*. Third World Network, 21 June, Geneva.

Greaves, T. 1996. Tribal rights. In *Valuing Local Knowledge: Indigenous People and Intellectual Property*, S. Brush and D. Stabinsky (eds.). Washington, D.C.: Island Press.

Jaff, W. and J. van Wijk. 1995. *The impact of plant breeders' rights in developing countries*. Amsterdam: IICA-University of Amsterdam. Draft.

Jullien, E. 1989. *Les impacts economiques de la protection de l'innovation sur le secteur european de la semence*. Paris: CERNA.

Kelsen, H. 1991. *Teoria Pura del Derecho*. Mexico, D.F.: Ed. Porrda.

Montecinos, C. 1996. *Sui generis*. A dead end alley, *Seedling* 13(4):19–28.

National Research Council. 1993. *Managing Global Genetic Resources: Agricultural Crop Issues and Policies*. Washington, D.C.: National Academy Press.

Nelson, R. 1971. La economia sencilla de la investigación scientifica basica. In *Selección, Económica del Cambio Tecnológico*, Nathan Rosenberg (ed.). Mexico, D.F.: Fondo de Cultura Económica.

Nijar, G.S. 1996a. *In Defense of Local Community Knowledge and Biodiversity*, Paper 1, Penang: Third World Network.

Nijar, G.S. 1996b. *TRIPs and Biodiversity. The Threat and Responsees: A Third World View*. Penang: Third World Network.

Oddi, A.S. 1996. TRIPs-natural rights and a polite form of economic imperialism, *Vanderbilt Journal of Transnational Law* 29(3):415–470.

Penrose, E. 1974. *La Economía del Sistema Internacional de Patentes*. Mexico, D.F.: Siglo Veintiuno Editores S.A.

Posey, D.A. and G. Dutfield. 1996. *Beyond Intellectual Property: Toward Traditional Resource Rights for Indigenous Peoples and Local Communities.* Ottawa: International Development Research Centre.

Shiva, V. 1995. *Captive Minds, Captive Lives: Ethics, Ecology and Patents on Life.* Dehra Dun: Research Foundation for Science, Technology and Natural Resource Policy.

Shiva, V. 1996. *Protecting Our Biological and Intellectual Heritage in an Age of Biopiracy.* Penang: Third World Network.

Swanson, T. 1995. Appropriation of evolution values. In *Intellectual Property Rights and Biodiversity Conservation,* T. Swanson (ed.). New York: Cambridge University Press.

Swanson, T., D. Pearce, and R. Cervigni. 1994. *The appropriation of the benefits of plant genetic resources for agriculture: an economic analysis of the alternative mechanisms for biodiversity conservation.* Report to the Commission on Plant Genetic Resources. Rome: FAO.

Tilahun, S. and S. Edwards (eds.). 1996. *The movement for collective intellectual rights.* Addis Ababa: The Institute for Sustainable Foundation/The Gaia Foundation.

United Nations Environmental Programme (UNEP). 1996. Knowledge, innovations and practices of indigenous communities: implementation of Article 8(j). Note by the Executive Secretary. Conference of the Parties to the Convention on Biological Diversity. 3rd Meeting 4–15 November 1996, Buenos Aires, Argentina. Available from <http://biodiv.org/cop3/cop3-19_vfinal.htm>.

Walden, I. 1995. Preserving biodiversity: the role of property rights. In *Intellectual Property Rights and Biodiversity Conservation,* T. Swanson (ed.). New York: Cambridge University Press.

Wegner, H. 1994. *Patent Law in Biotechnology, Chemicals and Pharmaceuticals.* Basingstoke, U.K.: Stockton.

chapter eleven

Farmer decision making and genetic diversity: linking multidisciplinary research to implementation on-farm

Devra Jarvis and Toby Hodgkin

Introduction

On-farm conservation has been proposed as a strategy to conserve the processes of evolution and adaptation of crops to their environments (Oldfield and Alcorn 1987; Altieri and Merrick 1987; Brush 1991). The conservation of specific genes or genotypes is secondary to the continuation of the processes that allow the material to evolve and change over time, remaining adapted to local agricultural production conditions. A prerequisite, however, for evolution and adaptation is the existence of genetic variation (Lande and Barrowclough 1990; Hamrick and Godt 1997). If the continued use of local cultivars by farmers is to form part of a conservation strategy, some knowledge of the amount of this genetic variation is needed to evaluate different approaches. This knowledge needs to be linked to farmer decision making and acquired over time (Frankel et al. 1995).

In the process of planting, managing, harvesting, and processing their crops, farmers make decisions that affect the genetic diversity of the crop populations. Over time they will modify the genetic structure of a population by selecting for plants with preferred agro-morphological characteristics. Farmers will influence the survival of certain genotypes by choosing a particular farming management practice or by planting a crop population in a site with a particular micro-environment. Farmers make decisions on the size of the population of each crop variety to plant each year, the percentage of seed to save from their own stock, and the percentage to buy or exchange from other sources. Each of these decisions, which can affect the genetic diversity of cultivars, is linked to a complex set of environmental and socio-economic influences on the farmer.

To date, the majority of on-farm conservation case studies have concentrated on linking farmer maintenance of local crop cultivars to environmental and socioeconomic factors at a particular point in time (Glass and Thurston 1978; Clawson 1985; Richards 1986; Brush 1991, 1995; Brush et al. 1992; Bellon 1996; Cromwell and van Oosterhout, this volume). These studies focus on investigating the factors that have influenced farmers to maintain or not to maintain local cultivars. In some cases, the genetic diversity of the locally grown cultivars has also been measured at a given time using genetic diversity indices of allelic richness or allele evenness within the population (e.g., Zimmerer and Douches 1991), but the primary concern of the investigation has usually been to describe the circumstances in which local cultivars constitute a part of production systems.

For the most part, the link between the effect of farmer management decisions and amount of genetic variation within the crop population has not been studied in detail. What is the effect of different farmer selection strategies on the genetic structure of the local cultivar over a number of years? What happens to the genetic diversity of local cultivars when farmers change the area planted? At what point will reduction in the area planted to a specific local cultivar lead to a significant reduction in the genetic diversity and limit further change? What is the effect on the genetic diversity of local cultivars of introducing new material or altering selection strategies through participatory breeding? These questions are important for understanding changing patterns of production and for those who advocate the use of local cultivars as components of sustainable production. They are also important questions for those who see on-farm maintenance of local cultivars as a component of a national conservation strategy. Without some understanding of the effect of farm-based decisions on genetic variation, national programs will lack the information needed to support, assist, or intervene in on-farm management of local cultivars where they see this as a part of their own conservation program. The few studies that have begun to look at the possibilities of a link between farmer decisions and genetic diversity have necessarily concentrated on one crop in one geographic area and focus on a particular point in time (Teshome 1996; Casas and Caballero 1996; Louette, this volume). For a national program to formulate a comprehensive on-farm conservation strategy, for each relevant within-country farming system, answers are needed to the questions: (1) *which farmer-based decisions affect whether the amount of genetic diversity within a crop population decreases, increases, or remains stable over time* and (2) *where, when, and how do these decisions affect the genetic diversity within a crop population?*

This chapter explores the issues involved in conserving genetic diversity through farmer maintenance and use of local cultivars. The concern is to consider the ways in which national plant genetic resources programs might address these issues. To illustrate this, we first describe the recently initiated program of work coordinated by the International Plant Genetic Resources Institute (IPGRI) which takes the form of a multidisciplinary global project to investigate on-farm conservation; to build national capacity and explore community participation in nine countries. We then discuss some central genetic questions and explore some of the implications of these for national programs concerned to maximize the diversity conserved.

Linking institutions, disciplines, and methodologies

In 1995, IPGRI, working through national programs, formulated a global project in nine countries to strengthen the scientific basis of *in situ* conservation of agricultural biodiversity. The objectives of the project are to:

- support the development of a framework of knowledge on farmer decision-making processes that influence *in situ* conservation of agricultural biodiversity
- strengthen national institutions for the planning and implementation of conservation programs for agricultural biodiversity
- broaden the use of agricultural biodiversity and the participation in its conservation by farming communities and other groups

The nine countries involved in the project are Burkina Faso, Ethiopia, Nepal, Vietnam, Peru, Mexico, Morocco, Turkey, and Hungary. In each country, strengthening the relations of formal institutions with farmers and local-level institutions to promote on-farm conservation is a major concern. These partner countries were included because each was within a region of primary diversity for crop genetic resources with worldwide importance. Each has traditional farming communities that maintain plant genetic resources. The countries all have national programs organized to conserve crop resources, which include *ex situ* conservation facilities, and all indicate a strong interest in developing a national capacity to support *in situ* conservation.

The program was formulated with the idea that *in situ* conservation activities should not aim to dissuade farmers from adopting new crop varieties that increase food availability or income, but rather to (1) determine and understand the situations in which local cultivars are maintained by farmers; (2) identify the key factors which affect farmer decisions to maintain local cultivars; (3) understand how farmer decision making affects the amount of genetic variation within crop populations over time; and (4) find ways to assist the continued selection of local cultivars or cultivars that conserve local germplasm. The development of any support program is expected to vary substantially across and within the different countries. The project emphasizes, therefore, participatory and learning approaches rather than the development of a specific model. The concern is to understand what is happening rather than prescribe abstract solutions.

The program supports research on the biological and social bases of *in situ* conservation, including (1) collecting a basic data set that links farmer decision making on the selection and maintenance of crop cultivars to measurable indices of genetic diversity; (2) training national scientists in *in situ* conservation research; (3) identifying target areas for *in situ* conservation programs; and (4) building bridges between conservationists, farmers, agricultural development agencies, and policy makers.

Three main strategies are used in project implementation: first, multidisciplinary work in the areas of crop biology and social sciences that will create a framework of knowledge and lead to institutional strengthening, methodologies and guidelines for other *in situ* programs; second, community participatory breeding and agronomic work, including community and locally based conservation activities involving market development, farmer incentives, and community-based training, that will support sustainable agricultural development; and third, international coordination and scientific synthesis to create a global framework for supporting *in situ* conservation by farmers.

The project works through formal institutions (e.g., universities, agricultural research and conservation institutes, and stations within the ministries of agriculture, natural resources, and science and technology) to strengthen their relations with farmers and local level institutions (e.g., community-based organizations, farmer groups, and non-governmental organizations) to promote on-farm conservation. Although working through national programs may be faulted as a centralized, top-down approach, the strategies that come out of this approach are based on information that comes from the farmer.

National programs interested in on-farm conservation must cope with the objectives of (1) conserving processes which promote genetic diversity of crop resources, and (2) ensuring the improvement of living standards of the farmer. To formulate an on-farm conservation strategy, knowledge is needed on how farmer selection practices affect crop genetic diversity. This requires scientific expertise from a variety of sources. At the same time, to ensure the sustainability of an on-farm conservation program, the national program needs to understand how, when, and where a farmer continues to maintain genetic variation. Sustainable management and conservation will be most effective where the resources have concrete value in the present time, can be used to meet the needs of local communities, and will contribute to the development of the nation as a whole (IPGRI 1996). The latter information requires contact with community-based organizations, extension workers, and non-governmental organizations that work closely with farmers. Such informal organizations are also in a position to recommend strategies that will influence a farmer to continue a particular selection practice that increases the amount of genetic variation in a population or to discontinue a selection process that decreases the amount of genetic variation in a population.

It might be argued that national program support of on-farm conservation is more likely to ensure its sustainability than non-governmental organization–funded projects. Funding to non-governmental organizations from donor agencies is finite, whereas, unless a change in government occurs, government funds can continue to be allocated. However, national institutes are far more removed from farming communities than non-governmental or community-based organizations. Therefore, the key is to ensure that a national program bases its on-farm conservation strategy on farmer-based perspectives by integrating the national program with community-based or non-governmental organizations. This is a difficult task as in many countries both parties are suspicious of each other. One role an international agency can play is to support projects that bring these different partners together.

Creating a framework of knowledge to support the formulation of on-farm conservation requires expertise from a variety of fields that are not normally associated: population genetics, biogeography, conservation biology, ecology, economics, sociology, anthropology, and local or "indigenous" knowledge systems. This expertise and the associated disciplines and methodologies are usually not found in a single institute. Often, a country's agricultural research institute may lack expertise in the social sciences. A social science department of a university may lack expertise in working directly with farmer groups. Community-based organizations may have expertise in working with farmers, but not in systematic sampling and relating farmer responses to population genetics. A meaningful investigation of questions relating farmer decision making to genetic diversity maintenance will require an integrated team of disciplines from formal institutes, such as universities and national research institutes, and informal organizations, such as community-based groups including non-governmental organizations.

Implementation of on-farm conservation investigations presupposes that such an integrated framework at central and local levels already exists within the country's national program. In many countries, this is not the case and the creation of such a framework is a prerequisite to formulating on-farm conservation strategies. For the IPGRI-supported global *in situ* project, the first step was to support the formation of such a framework and integrated teams in the nine participating countries. This framework has consisted of the setting up of a multidisciplinary National Advising Committee, which is led by a National Project Coordinator and includes members from formal and informal institutes. The National Advising Committee serves as the lead institution in coordinating and monitoring project activities, provides technical backup, ensures integration into the national program and approves plans and reports for the regional and global management levels. In addition, technical working groups are established in biological and social sciences and in extension and training. These involve ethnic and gender groups and will technically supervise and monitor project activities. Technical support is supplied by a project Technical Advisory Committee. The way in which this is working out in practice is briefly illustrated for Morocco, Burkina Faso, and Nepal.

Morocco

Situated along the Atlantic Ocean and the Mediterranean Sea, with the Rif Mountains in the North, the Atlas Mountains running north to south, and the Sahara Desert east and south of the Atlas Mountains, Morocco contains a unique array of agroecosystems. It is a center of diversity for such world-wide crops as wheat, barley, faba bean, and alfalfa (Neal-Smith 1955; Nègre 1956; Perrino et al. 1984; Tazi et al. 1989). The country's crop diversity results from long-term adaptation to drought, cold, and saline conditions (Sauvage 1975; Graves 1985; Birouk 1987; Francis 1987; Birouk et al. 1991). Islands of crop diversity with a high dependence on local cultivars remain in mountain areas and in oases at the edge of the Sahara.

Morocco has a national program for plant genetic resource (PGR) conservation coordinated by a National Committee, which was established in 1992. The management committee for the program consists of members from nine different organizations and institutes. The National Project Committee for the on-farm conservation project in Morocco is a subcommittee of this National Committee. Formal institutes involved are Hassan II Institute of Agronomy and Veterinary Medicide (IAV) and Institut National de la Recherche Agronomique (INRA) and their associated extension and research institutes, the Agricultural Provincial Directorates (DPA) and Office Regional de Mise en Valeur Agricole (ORMVA). Hassan II IAV is involved in research in cultivated cereals and fodder crops, *ex situ* conservation, evaluation, genetic analysis, GIS, documentation and data analysis, socioeconomic studies, and ethnobotany. INRA is responsible for national agricultural research and includes departments for Cereals, Legumes, Soil Science and the Environment, Socioeconomics, and Genetics. The DPA and ORMVA, under the Ministry of Agriculture, are involved with local farmer contact, extension, and technology transfer. A representative from the Education Department of the Ministry of Agriculture is involved in the project for public awareness and extension work at the central level. In addition, L'Association pour la Preservation de la Biodiversité au Maroc (BELDIA), a Moroccan non-governmental organization versed in community participatory projects, is a member of the National Project Committee.

Three priority agro-ecological zones within Morocco have been selected for the project. The first is the Demnate/Tanante region, under the Azilal DPA in the high Atlas, a semi-arid area with a clay loam soil. Priority crops for this area are barley, durum wheat, faba bean, and alfalfa. The second region is the Valley of Ziz-Fafilalte, managed by the Errachidia ORMVA in the oasis area, an area of semi-desert with sandy loam soils. The priority crops here are bread wheat and alfalfa. The final area is under Taza DPA and Chefchaouen DPA in the Rif and Pre-Rif mountain area. Here barley, durum wheat, and faba bean are grown under rain-fed conditions, on a clay loam soil.

Burkina Faso

Burkina Faso sits in the Soudanian and Sahelian zones and is within the region of African crop domestication, evolution, and diversity for sorghum, millet (especially pearl millet), and cowpea. Sorghum, pearl millet, and, more recently, maize and rice occupy more than 75% of the cultivated land, while groundnut and cowpea occupy 25%. Sorrel, onion, and okra are important as vegetables in the production of sauces and as cash crops. Approximately 75% of the country lies on old Precambrian crystalline rock, with extensive areas of marginal soils. Rainfall is extremely variable, and the amount and length of season of rainfall are main factors affecting crop yields. Serious soil degradation, repeated drought, and unrestrained use of modern varieties contribute to the erosion of genetic diversity, but, in some areas, local cultivars are grown extensively.

In Burkina Faso, the Centre National de la Recherche Scientifique et Technologique (CNRST) houses key institutes and scientists who work in the area of plant genetic resource conservation and use. A national strategic plan has been developed for research, which includes on-farm conservation of plant genetic resources. CNRST has capacity both for research and to make links to farmers via protocols, which it has already signed with a number of non-governmental organizations. Two institutes are especially important for investigating the scientific basis of on-farm conservation: the Institut d'Études et Recherches Agricoles (INERA) and the Institut de Recherche en Sciences Sociales et Humaines (IRSSH). The Institut de Recherches en Biologie et Ecologie Tropicale (IRBET), focusing on ecology, will soon be combined with INERA.

INERA has strong link to the national non-governmental organization, the Fédération de Unions des Groupements Naam (FUGN), a farmers' union that operates throughout Burkina Faso. In addition, two other non-governmental organizations, CRPA-Yatenga and Crocevia, operate a program on production of seeds of local varieties including African rice, sorghum, and millet. Another important member of the National Project Committee is the Université de Ouagadougou, which has both research facilities and training capacity.

Agroecological regions were selected from the three major climatic zones of Burkina Faso: the Sahelian, the North Soudanian, and the Soudanian. Research sites were also chosen for degree of population density, ranging from the more densely populated region in central Burkina Faso to the less densely populated areas in the north and southwest. In the Sahelian zone conditions are harsh: precipitation is less than 600 mm per year and soils are poorly developed. The region has a well-organized agricultural extension network in place, together with nongovernmental organization presence and farmers' organizations. In the north Soudanian zone precipitation ranges from 600 to 800 mm, population density is extremely high, and there are many farmers' organizations interested in participating in the project. The Soudanian zone selected contains the major food production area of Burkina Faso. Here the precipitation is greater than 800 mm. Soils are lateritic or hydromorphic (along the Mouhoun River), and the region is served by a strong extension program with active nongovernmental and farmer organizations.

Nepal

Nepal's location, geography, and diverse ethnic groups have made it an important center of agricultural biodiversity. Crop production takes place between 70 to 3000 m elevation, under rainfall regimes from less than 1000 mm to over 5000 mm, and on infertile to very fertile soils. Numerous ethnic groups of both Indo-Arayan and Tibet-Burmese descent, with varying cultural preferences, and different access to market, credit, and agricultural inputs, have created a diversity of agroecological systems (Upadhyay and Sthapit 1995). Food grains, such as rice and finger millet, grain legumes, barley, and minor crops, including buckwheat, have great genetic diversity. More than half of the arable land is planted to local cultivars. However, with the introduction of new high yielding varieties and over-exploitation of natural resources, genetic erosion is increasing rapidly in Nepal.

The Nepal Agricultural Research Council (NARC) is the national focal point for agricultural research in Nepal. Recently NARC has begun working with farmers through a network of regional research centers to further participation in *in situ* activities. Agricultural research stations such as Jumla, Lumle, Malepatan (Pokhara and Hill Crop Program at Kavre), and Parwanipur have been identified as potential partners within NARC systems for *in situ* activities, together with local non-governmental organizations with relevant experience (e.g., Local Initiatives for Biodiversity Research and Development, LI-BIRD). The Chief of the Agricultural Botany Division is the National Project Coordinator for the Nepal project and the National Project Committee is comprised of members from NARC, the Ministry of Agriculture, and non-governmental organizations.

Three regions of Nepal have been selected to represent high, middle, and low altitudes of crop production ecosystems in Nepal (Upadhyay and Sthapit 1995). The Jumla valley is remote and has a unique range of crop varieties finely adapted to local conditions. The area is a transition zone between lower elevations, where a winter cereal is followed by a summer crop, and higher elevations where only one crop (either winter cereal or summer crop) can be obtained. The valley is known for its cold-tolerant Jumli Marshi rice and other crops associated with rice-based farming systems. The second area, the Pokhara Seti River Valley, is known for its quality rice in the Western Hills of Nepal. It is characterized by a number of lakes, broad alluvial valleys, isolated hills, and meandering streams. Rice-wheat-vegetables and maize-millet-vegetables are major cropping systems. The valley has diverse ethnic composition, mainly Brahmin, Chhetri, Gurung, Magar, and Newar. Parwanipur, the third area, lies in the fertile strip of Indo-Gangetic plain (100 to 200 m) on the southern frontier bordering with India. The production potential is high and farmers have adequate access to inputs. The rice-wheat system is the basic cropping pattern of the region and both irrigated and rain-fed systems occur in the same communities.

Implementing integrated research on-farm

The frameworks described above for Morocco, Burkina Faso, and Nepal are similar to those envisaged for the other partner countries in the IPGRI-supported, global on-farm conservation program. Once a project management framework is in place, an initial step for each partner country has been to select regions for the work. Initial agro-ecological identification is followed by a natural and social science baseline survey carried out by a multidisciplinary team as a preliminary to specific site selection. Multidisciplinary teams are needed at the very start to evaluate if the initial agroecological zones selected meet mutually agreed-upon criteria, such as the existence of genetic diversity, desired agroecological variation, accessibility to the locality, links to agricultural extension work, and, most importantly, interest and cooperation of local communities. In Nepal, for example, multidisciplinary teams have been formed at the national level and are planned for each agroecological site. The national level team consists of agroecosite managers, one from each of the three agroecosites, a crop biologist, a social scientist, an ecologist, a gender specialist, an outreach specialist, and a participatory plant breeding specialist. Similarly, local teams are planned for each of the three agroecological sites. In Morocco, the team consists of specialists from the national research system and Hassan II University in each of the priority crops, soil science, socioeconomics, and genetics, together with staff from the national agricultural extension and outreach program and the nongovernmental organization, BELDIA. In each country, once members of each team are identified at the national and local level, the actual selection of project sites and farmer participants becomes an interactive process between researchers, agricultural extension workers, and the farm community.

Preliminary to baseline data collection and site selection, training of local extension agents/non-governmental organizations and scientific research workers is needed in participatory approaches, semi-structured interviews, identification and use of key informants, gender sensitivity, and other aspects of gathering socioanthropogenic information. Likewise, both social and natural scientists participating in the research should receive some basic orientation on population genetic concepts, and on the ecological data required. In Nepal and Burkina Faso, workshops are being organized for geneticists, ecologists, social scientists, and community-based staff, as well as with community representatives and participating farmers, to develop an understanding of the data needed and the work envisaged.

The objective of the information-gathering component of the project is to develop a set of data that can be used to answer questions about on-farm maintenance as a conservation process. It is designed to explore the links between environmental features, farmer decision making, and the genetic diversity maintained over time in local cultivars. One key area on which it is hoped to obtain information is the link between farmer-based decisions and the extent and distribution of genetic diversity in local cultivars in the study areas. Five aspects of farmer decision making seem likely to be of most importance: (1) decisions on what agro-morphological characteristics are important to the farmer in any selection procedures used; (2) decisions on what farming practice to use on the local cultivar population; (3) decisions on where to plant the population; (4) decisions on the size of the population to plant; and (5) decisions on the seed source for the population. It is hoped to establish the effect of these decisions on the genetic diversity of the different populations studied over time.

Agro-morphological characteristics

Local cultivars are normally defined by farmers in terms of their agro-morphological characteristics (Zimmerer and Douches 1991; Weltzien et al. 1996; Sthapit et al. 1996b; Louette and Smale 1996; Teshome 1996; Louette et al. 1997). Depending on the crop, a farmer may decide to select and maintain plants based on preferred agro-morphological criteria, such as early flowering, height, denseness of inflorescence, or a particular color, shape, or taste (Boster 1985). Some of these characteristics may be controlled by single genes but most are controlled by many loci, as in the case of most yield-related characters. A first set of questions concerns the way in which character or performance based selection by farmers influences the overall genetic diversity of the population as well as that of the characters of concern to the farmer. Farmer-based selection, even experimentation (Richards 1986), is often very important but may not always be so (Ceccarelli and Grando, Chapter 3, this volume), and may be concerned primarily with maintenance rather than change (Bellon et al. 1997). In any integrated studies, on-site observations will also be needed to investigate whether there is introgression between crops and their wild relatives and whether these are noted and retained (or discarded) by farmers (Jarvis and Hodgkin 1996). In Burkina Faso, the national program is currently conducting studies on the natural introgression of wild and cultivated pearl millet.

A second set of questions relevant to farmer selection of preferred agro-morphological characters is: does the farmer-based selection process change over time in accordance with changes in environmental or socioeconomic conditions, and, if so, under what conditions? Where changes in the selection process occur, the extent of any associated changes in the genetic diversity of the population over time needs to be investigated.

The approach proposed in the IPGRI global project is to first ask farmers to list the agro-morphological characteristics used to distinguish a crop variety, and then to prioritize the characteristics he or she selects each year. Selection of plants or seeds with priority agro-morphological characteristics may be made in the field throughout the growing season or after harvest. The selector may be male or female, young or old. Collection of selection criteria, therefore, may need to be acquired separately for different gender and age groups for it to be meaningful. Based on the above information, researchers can then select, sample, and measure within populations of each selected variety over time and space: (1) those characteristics prioritized by farmers that are known to be heritable and easily measurable; (2) other heritable agro-morphological characteristics that are not purposefully selected for by the farmer; and (3) selected biochemical and molecular markers, depending on the capacity of the national program carrying out the research.

Farming management practices

Farming management practices include land preparation, planting, thinning and weeding, fertilizer application, pest control, irrigation, harvesting, and post-harvest processing. Each of the processes may or may not affect the amount of genetic diversity of the crop population over time (Snaydon 1984). Different levels of fertilizer application or the use of organic rather than chemical fertilizer may select for different genotypes in a population (Silvertown et al. 1994). Irrigated and non-irrigated populations of faba beans in Morocco have different population genetic characteristics (Sadiki 1990). Dense planting to reduce weeding and post-harvest storage of seeds also play a selection role on seed survival and the continuation of subsequent traits in the next generation.

Key questions for national on-farm conservation programs are (1) which farming management practices influence genetic diversity and (2) to what extent do these practices affect the amount of genetic variation in the crop population. For national programs to look at these questions, interviews and observations of farmer management practices are needed. Field trials using different degrees of density or fertilizer application coupled with measurements of genetic diversity may be desirable to investigate specific effects of particular treatments. Such studies may, however, lie outside the capacity of this particular project and be best explored in specific case studies.

Environmental selection

When a farmer decides to plant a cultivar on a particular micro-site, the crop population is exposed to specific environmental selection. By planting barley in soils prone to water logging or in different agroecological sites, the Ethiopian farmer is subjecting the plants to environmental selection for tolerance to a specific stress (Demissie and Bjornstad 1996, 1997). Similarly, by planting upland "Jumla" rice at elevations up to 3000 m in Nepal, or "Chao" rice on shallow soils in northern Vietnam, the farmer is exposing a crop population to cold stress or poor soil conditions.

Questions under this selection category revolve around (1) how environmental selection has influenced the genetic diversity of the population over time and (2) which environmental conditions play a significant role in affecting the amount of genetic variation in the crop population. The farmer's involvement is in making the decision to plant that particular population in a particular micro-habitat.

To answer these questions, national programs will need to focus on basic questions of ecological genetics, to understand how populations adapt to their environments, and to determine the type and extent of effect of environmental factors on the amount of genetic variation over time (Merrell 1981; Allard 1988, 1990; Fowler 1990; Real 1994; Le Boulc'h et al. 1994; Goldringer et al. 1994). The extent to which farmers maintain unique types for particular micro-environments will be important, as well as the occurrence within local cultivars of G × E interactions or broad tolerance of a range of stress environments (Huenneke 1991; Via 1994; Anikster et al. 1997). Social scientists' involvement will be in investigating what factors influence a farmer to plant in the particular habitat (Bellon and Taylor 1993; Bellon and Brush 1994). The work will likely point to specific studies over longer periods that will help breeders and users understand the nature of adaptation in local cultivars.

Population size

The size of a population planted by a farmer will affect the amount of genetic variation of the crop population over time (Shaffer 1990; Lande and Barrowclough 1990; Barrett and Kohn 1991). The smaller the population, the more likely it is that genetic drift, inbreeding, loss of alleles, and stochastic events will affect the population (Shaffer 1990; Frankel et al. 1995; Slatkin 1987, 1994). From a conservation perspective, crucial questions include: How does population size influence the genetic diversity of the local cultivar population? What is the effective population size that ensures long-term stability of the population? Should a national program be concerned with a group of smaller populations or metapopulations, or should the conservation focus be on individual farmer fields? To understand the effect of size on the amount of genetic variation of a population requires minimum population viability studies and population genetic methods (Menges 1991; Caballero 1994).

Seed source

A farmer makes a choice each year on what percentage of his or her own seeds to save and plant and what percentage he or she will acquire from other farmers. How do migration (influx of new seeds) and bottlenecks (reduction in the number of saved seeds) affect the genetic diversity of the population over time? Louette and Smale (1996) have shown that after six crop rotations of the farmer studies in Mexico, only 48% of the seed material remained from the farmer's original stock. This is the information that a population geneticists will need to determine effective population sizes (Levin 1984; Louette et al. 1997).

Again, the question of whether a metapopulation should be the unit of concern for national conservation programs is important (Henry et al. 1991;

Louette, this volume). This involves social scientists and community-based groups developing an understanding of the seed supply system in order for plant geneticists to be able to determine the effective population size for the cultivar in question (Friis-Hansen 1996; Cromwell 1996). This requires knowledge, not only of seed source and informal seed supply systems, but also of how different storage systems influence the survival of different seeds over time (Kashyap and Duhan 1994).

On-farm data for conservation and use

On-farm maintenance of local cultivars is currently being promoted as a strategy to conserve genetic diversity and as a process that will secure the continued availability of genetic variation while ensuring the continuing betterment of farmer livelihood over time. In the past, most conservation workers have focused largely on *ex situ* conservation, with the expectation that local cultivars maintained over the centuries will shortly disappear. This is now seen as an oversimplification of the complex issues at play (Louette, this volume). Knowledge of the processes involved in farmer selection and maintenance of crop cultivars can be used by national programs to make decisions on where to support, assist, or intervene to promote the conservation of genetic diversity and strengthen the effectiveness of conservation. How can this same information, linking farmer decision making to genetic diversity, be used to help national programs improve the livelihood of the farmer? To be maintained by farmers, crop genetic resources must have value to them and must retain this value and be competitive to other options a farmer might have, or even better, to increase this value such that it is reflected in an increase in a farmer's standard of living.

Value may be added to crop resources in two main ways: the material itself may be improved or the demand for the material or some product may increase. One option is to seek improved quality, disease resistance, yield, taste, or other farmer preferred characteristics, through decentralized breeding activities (Ceccarelli et al. 1996; Eyzaguirre and Iwanaga 1996; Joshi and Witcombe 1996; Sthapit et al. 1996a, Sthapit et al. 1996b; Witcombe et al. 1996). Sthapit, Joshi, and Witcombe (1996a) have shown that, by utilizing farmers' knowledge, acceptable varieties can be bred with minimum use of resources in the high altitude areas of Nepal. Effective support for farm-based maintenance needs to explore the ways in which this can best be done and the impact it has on the extent and distribution of genetic diversity.

Value can also be added to crop resources by better processing, storage, and marketing, where the farmer receives more benefit from the final product. An important role for community-based organizations and non-governmental organizations is the formation of farmer cooperatives and farmer-managed community seed banks to maximize returns to the farmers themselves (Worede 1992, 1997; Gaifami 1992). Government policy may also play a role in ensuring that farmers' inputs and seeds receive the same or better market treatment and government support as improved varieties (Leskien and Flitner 1997; Qualset et al. 1997).

Conclusion

The majority of early studies on on-farm conservation have focused on trying to determine what factors have caused a farmer to maintain or not maintain a diversity of local cultivars. Other studies have looked at the amount of genetic variation of these cultivars at a given moment in time. The research discussed in this chapter aims to take earlier studies further by attempting to quantify over time how specific farmer-based decisions may determine both the amount of genetic variation and the effective population size of a cultivar population over time.

Changes to the amount of genetic variation can be measured as can the effect of farmer-based management on the effective population sizes. Genetic variation may be linked to farmer decisions in five major categories: agro-morphological characteristics, farm management practices, planting location, size of the population, and seed source. These decisions in turn are based on environmental and socioeconomic influences. The first link forms the basic data set for the second link. Understanding these processes requires the involvement of people from informal and formal institutes and from a variety of scientific disciplines. Once there is better knowledge of these processes, national programs will be in a better position to support, assist, or intervene in the conservation of genetic diversity on-farm.

Farmers themselves will ultimately determine the extent to which on-farm maintenance of local cultivars continues and, hence, contributes to the overall conservation of crop genetic diversity. Previous conservation work underplayed or even ignored this contribution and emphasized *ex situ* methods of diversity maintenance. Through the creation of a framework of knowledge of the processes involved, it is hoped to provide a more complete understanding that redresses this imbalance.

References

Allard, R.W. 1988. Genetic changes associated with the evolution of adaptedness in cultivated plants and their wild progenitors, *Journal of Heredity* 79:225–238.

Allard, R.W. 1990. The genetics of host-pathogen coevolution implications for genetic resources conservation, *Journal of Heredity* 81:1–6.

Altieri, M.A. and L.C. Merrick. 1987. *In situ* conservation of crop genetic resources through maintenance of traditional farming systems, *Economic Botany* 41:86–96.

Anikster, Y., M. Feldman, and A. Horovitz. 1997. The Ammiad experiment. In *Plant Genetic Conservation: The in situ Approach*, N. Maxted, B.V. Ford-Lloyd, and J.G. Hawkes (eds.). London: Chapman & Hall.

Awegechew, T. 1996. Factors Maintaining Sorghum [Sorghum Bicolor (L.) Moench] Landrace Diversity in North Shewa and South Welo Regions of Ethiopia. Ph.D. thesis, Ottawa: Carleton University.

Barrett, S.C.H. and J. R. Kohn. 1991. Genetic and evolutionary consequences of small population size in plants: implications for conservation. In *Genetics and Conservation of Rare Plants*, D.A. Falk and K.E. Holsinger (eds.). New York: Oxford University Press.

Bellon, M.R. 1996. The dynamics of crop infraspecific diversity: a conceptual framework at the farmer level, *Economic Botany* 50:26–39.

Bellon, M.R. and S.B. Brush. 1994. Keepers of maize in Chiapas, Mexico, *Economic Botany* 48:196–209.

Bellon, M.R. and J.E. Taylor. 1993. Farmer soil taxonomy and technology adoption, *Economic Development and Cultural Change* 41:764-786.

Bellon, M.R., J.-L. Pham, and M.T. Jackson. 1997. Genetic conservation: a role for rice farmers. In *Plant Genetic Conservation: The in situ Approach*, N. Maxted, B.V. Ford-Lloyd, and J.G. Hawkes (eds.). London: Chapman & Hall.

Birouk, A. 1987. *Les ressources phytogénétiques au Maroc: analyse de la variabilité génétique d'une espèce fourragere: la luzerne* (Medicago sativa L.). Ph.D. Thesis, IAV Hassan II.

Birouk, A., J. Lewalle, and M. Tazi. 1991. Le patrimoine végétal des provinces sahariennes du Maroc. Documents Scientifiques et Technique, Actes Editions, Institute Agronomique et Veterinaire Hassan.

Boster, J.S. 1985. Selection for perceptual distinctiveness: evidence from Aguaruna cultivars of *Manihot esculenta*, *Economic Botany* 39:310–325.

Brush, S.B. 1991. A farmer-based approach to conserving crop germplasm, *Economic Botany* 45:153–65.

Brush, S.B. 1995. *In situ* conservation of landraces in centers of crop diversity, *Crop Science* 35:346–54.

Brush, S.B., J.E. Taylor, and M.R. Bellon. 1992. Biological diversity and technology adoption in the Andean potato agriculture, *Journal of Development Economics* 39:365–387.

Caballero, J. 1994. Developments in the prediction of effective population size, *Heredity* 73:657–679.

Casas, A. and J. Caballero. 1996. Traditional management and morphological variation in *Leucaena esculenta* (*Fabaceae: Mimosoideae*) in the Mixtec region of Guerrero, Mexico, *Economic Botany* 50:167–181.

Ceccarelli, S., S. Grando, and R.H. Booth. 1996. International breeding programs and resource-poor farmers: Crop improvement in difficult environments. In *Participatory plant breeding. Proceedings of a workshop on participatory plant breeding, 26–29 July 1995, Wageningen, The Netherlands*, P. Eyzaguirre and M. Iwanaga (eds.). Rome: IPGRI.

Clawson, D.L. 1985. Harvest security and intraspecific diversity in traditional tropical agriculture, *Economic Botany* 39:56–67.

Cromwell, E. 1996. *The Seed Sector in Perspective. Governments, Farmers and Seeds in a Changing Africa*. Wallingford, Oxon, U.K.: CAB International.

Demissie, A. and Bjornstad, A. 1996. Phenotypic diversity of Ethiopian barleys in relation to geographical regions, altitudinal range, and agro-ecological zones: as an aid to germplasm collection and conservation strategy, *Hereditas* 124:17–29.

Demissie, A. and Bjornstad, A. 1997. Geographic, altitude and agro-ecological differentiation of isozyme and hordein genotypes of landrace barleys from Ethiopia: implications to germplasm conservation, *Genetic Resources and Crop Evolution* 44:43–55.

Eyzaguirre, P. and M. Iwanaga. 1996. Farmers' contribution to maintaining genetic diversity in crops, and its role within the total genetic resources system. In *Participatory plant breeding. Proceedings of a workshop on participatory plant breeding, 26–29 July 1995, Wageningen, The Netherlands*, P. Eyzaguirre and M. Iwanaga (eds.). Rome: IPGRI.

Fowler, N.L. 1990. The Effects of competition and environmental heterogeneity on three coexisting grasses, *Journal of Ecology* 78:389–402.

Francis, C.M. 1987. *Morocco, a plant collection tour*. A report compiled for the Grain Research Committee of Western Australia, Western Australian Department of Agriculture, Perth.

Frankel, O., A.D.H. Brown, and J.J. Burdon. 1995. *The Conservation of Plant Biodiversity*. Cambridge: Cambridge University Press.

Friis-Hansen, E. 1996. The role of local plant genetic resource management in participatory breeding. In *Participatory plant breeding. Proceedings of a workshop on participatory plant breeding, 26–29 July 1995, Wageningen, The Netherlands*, P. Eyzaguirre and M. Iwanaga (eds.). Rome: IPGRI.

Gaifami, A. 1992. Developing local seed production in Mozambique. In *Growing Diversity*, D. Cooper, R. Vellve, and H. Hobbelink (eds.). London: Intermediate Technology Publications.

Glass, E.H. and H.D. Thurston. 1978. Traditional and modern crop protection in perspective, *Bioscience* 28:109-115.

Goldringer, I. J.-L. Pham, J.L. David, P. Brant, and A. Gallais. 1994. Is dynamic management of genetic resources a way of pre-breeding? In *Evaluation and Exploitation of Genetic Resources Pre-Breeding. Proceedings of the Genetic Resources Section Meeting of Eucarpia*, F. Balfourier and M.R. Perretant (eds.). 15–18 March. Clermont-Ferrand, France.

Graves, W. 1985. *Moroccan indigenous plants collection program*. TDY Report, Utah State University, Range Improvement Project, Rabat, Morocco.

Hamrick, J.L and J.W. Godt. 1997. Allozyme diversity in cultivated plants, *Crop Science* 37:26–30.

Henry, J.P., C. Pontis, J. David, and P.H. Gouyon. 1991. An experiment on dynamic conservation of genetic resources with metapopulations. In *Species Conservation. A Population-Biological Approach*, A. Seitz and V. Loeschcke (eds.). Basel: Birkhauser Verlag.

Huenneke, L. 1991. Ecological implications of genetic variation in plant populations. In *Genetics and Conservation of Rare Plants*, D. Falk and K. Holsinger (eds.). Oxford: Oxford University Press.

IPGRI. 1996. *An IPGRI strategy for the* in situ *Conservation of Agricultural Biodiversity*. Rome: IPGRI. Unpublished.

Jarvis, D. and T. Hodgkin. 1996. Wild relatives and crop cultivars: conserving the connection. Paper presented at the International Symposium on *in Situ* Conservation of Plant Genetic Resources, Antalya, Turkey, November 1996.

Joshi, A. and J.R. Witcombe. 1996. Farmer participatory crop improvement. II. Participatory varietal selection, a case study in India, *Experimental Agriculture* 32:461–477.

Kashyap, R.K. and Duhan, J.C. 1994. Health status of farmers' saved wheat seed in Harayana, India — A case study, *Seed Science and Technology* 22:619–628.

Lande, R. and G. Barrowclough. 1990. Effective population size, genetic variation, and their use in population management, *Viable Populations for Conservation*, M. Soulé (ed.). Cambridge: Cambridge University Press.

Le Boulc'h, V., J.L. David, P. Brabant, and C. De Vallavieille-Pope. 1994. Dynamic conservation of variability: responses of wheat populations to different selective forces including powdery mildew, *Genetics Selection Evolution* 26, Suppl. 1:221s–240s.

Leskien D. and M. Flitner. 1997. Intellectual property rights and plant genetic resources: options for a sui generis system. *Issues in Genetic Resources*. No. 6. Rome: IPGRI.

Levin, D.A. 1984. Immigration in plants: an exercise in the subjunctive. In *Perspectives on Plant Population Ecology*, R. Dirzo and J. Sarukhan (eds.). Sunderland, MA: Sinauer Associates, Inc.

Louette, D. and M. Smale. 1996. *Genetic Diversity and Maize Seed Management in a Traditional Mexican Community: Implications for in Situ Conservation of Maize*. NRC Paper 96-03. Mexico, D.F: CIMMYT.

Louette, D., A. Charrier, and J. Berthaud. 1997. *In situ* conservation of maize in mexico: genetic diversity and maize seed management in a traditional community, *Economic Botany* 51:20–38.

Menges, E.S. 1991. The application of minimum viable population theory to plants. In *Genetics and Conservation of Rare Plants*, D.A. Falk and K.E. Holsinger (eds.). New York: Oxford University Press.

Merrell, D.J. 1981. *Ecological Genetics*. Minneapolis: University of Minnesota Press.

Neal-Smith, C.A. 1955. *Report on herbage plant exploration in the Mediterranean region*. F.A.O. Report 415. Rome: FAO.

Nègre, R. 1956. *Les luzernes du Maroc. Travaux de L'institute Scientifique Chérifien, Rabat*. Série Botanique 5. Rabat: Institute Scientifique Chérifien.

Oldfield, M.L. and J.B. Alcorn. 1987. Conservation of traditional agroecosystems, *Bioscience* 37:199–208.

Perrino P., G.P. Polignano, J.S. Kwong, and M. Khouya-Ali. 1984. Collecting germplasm in southern Morocco. *FAO/IBPGR Plant Genetic Resources Newsletter* 65:26–28.

Qualset, C.O., A.B. Damania, A.C.A. Zanatta, and S.B. Brush. 1997. Locally based crop plant conservation. In *Plant Genetic Conservation: The in situ Approach*, N. Maxted, B.V. Ford-Lloyd, and J.G. Hawkes (eds.). London: Chapman & Hall.

Real, L. 1994. Introduction: Current directions in ecological genetics. In *Ecological Genetics*, L. Real (ed.). Princeton, NJ: Princeton University Press.

Richards, P. 1986. *Coping with Hunger: Hazard and Experiment in an African Rice-Farming System*. London: Allen and Unwin.

Sadiki, M. 1990. *Germplasm development and breeding of improved biological nitrogen fixation of Faba bean in Morocco*, Ph.D. Dissertation. Minneapolis: University of Minnesota.

Sauvage, C. 1975. L'etat actuel de nos connaissances sur la flore du Maroc. *La Fore du Bassin Mediterraneen. Essai de Systématique, Synthétique*. Paris: CNRS.

Shaffer, M. 1990. Minimum viable populations: coping with uncertainty. In *Viable Populations for Conservation* (4th ed.), M.E. Soulé (ed.). Cambridge: Cambridge University Press.

Silvertown, J., D.A. Wells, M. Gilman, M.E. Dodd, et al. 1994. Short-term and long-term after effects of fertilizer application on the flowering population of green winged orchid *Orchis morio*, *Biological Conservation* 69:191–197.

Slatkin, M. 1987. Gene flow and the geographic structure of natural populations, *Science* 236:787–836.

Slatkin, M. 1994. Gene flow and population structure. In *Ecological Genetics*, L. Real (ed.). Princeton, NJ: Princeton University Press.

Snaydon, R.W. 1984. Plant demography in an agricultural context, *Perspectives on Plant Population Ecology.* Sunderland, MA: Sinauer Associates, Inc.

Sthapit, B.R., K.D. Joshi, and J.R. Witcombe. 1996a. Farmers' participatory high altitude rice breeding in Nepal: providing choice and utilizing farmers' expertise. In *Using Diversity: Enhancing and Maintaining Genetic Resources On-Farm*, M. Loevinsohn and L. Sperling (eds.). New Delhi: International Development Research Center for South Asia.

Sthapit, B.R., K.D. Joshi, and J.R. Witcombe. 1996b. Farmer participatory crop improvement. III. Participatory plant breeding, a case study for rice in Nepal. *Experimental Agriculture* 32:479–496.

Tazi, M., A. Birouk, Z. Fatemi, and P. Heiffer. 1989. Collecting germplasm in Morocco, *FAO/IBPGR Plant Genetic Resources Newsletter* 77:39.

Upadhyay, M.P. and B.R. Sthapit. 1995. Plant Genetic Resource Conservation Programs in Nepal: Some Proposals for Scientific Basis of *in situ* Conservation of Agro-biodiversity. Paper presented at the Participants' Conference on *In Situ* Conservation of Agricultural Biodiversity, hosted by IPGRI in Rome, 17–19 July 1995.

Via, S. 1994. The evolution of phenotypic plasticity: what do we really know? In *Ecological Genetics*, L. Real (ed.). Princeton, NJ: Princeton University Press.

Weltzien, E.R., M.L. Whitaker, and M.M. Anders. 1996. Farmer participation in pearl millet breeding for marginal environments. In *Participatory plant breeding. Proceedings of a workshop on participatory plant breeding, 26–29 July 1995, Wageningen, The Netherlands*, P. Eyzaguirre and M. Iwanaga (eds.). Rome: IPGRI.

Witcombe, J.R., K.D. Joshi, and B.R. Sthapit. 1996. Farmer participatory crop improvement. I: Varietal selection and breeding methods and their impact on biodiversity, *Experimental Agriculture* 32:445–460.

Worede, M. 1992. Ethiopia: a gene bank working with farmers. In *Growing Diversity*, D. Cooper, R. Vellve, and H. Hobbelink (eds.). London: Intermediate Technology Publications.

Worede, M. 1997. Ethiopian *in situ* conservation. In *Plant Genetic Conservation: The in situ Approach*, N. Maxted, B.V. Ford-Lloyd, and J.G. Hawkes (eds.). London: Chapman & Hall.

Zimmerer, K.S. and D.S. Douches. 1991. Geographical approaches to native crop research and conservation: the partitioning of allelic diversity in Andean potatoes, *Economic Botany* 45:176-189.

Index

A

ABA, *see* Asociación Bartolomé Aripaylla
Adaptation variations in landraces, 35–36
Agrarian Reform of 1950, 112
Agreement on Trade Related Aspects of
    Intellectual Property Rights
    (TRIPs), 23, 250, 259
Agricultural development
    changes in farming systems, 11–12
    diversity and poverty relationship, 7
    domestication progression, 5–6
    dynamics of local crop diversity, 33
    germplasm conservation efforts, 7
    impact of information on, 191
    impact of integration, 6
    optimal conservation and, 190–191
    promotion of *in situ* conservation, 7–8
Agricultural production function, 177
Allozyme diversity, 33–35
Andean Ecoregional Initiative, 202
Andean landrace study, *see* Culture of the
    Seed in Peruvian Andes
Andean Root and Tuber Crops Project, 202
Appellation/certification system, 19
Arabi Abiad (white seed barley), *see* Barley
    landraces in the Fertile Crescent
Arabi Aswad (black seed barley), *see* Barley
    landraces in the Fertile Crescent
Arachis hypogea, 36
Arta barley strain, 59–60
Asociación Bartolomé Aripaylla (ABA)
    first seed diversity fair, 208–209
    recovery project background, 20, 196, 210
    sample crops, 212, 213–215
    second seed diversity fair, 209–210
Ayllu, 203, 205
Aymara and Quechua communities, 205

B

Barley (*Hordeum vulgare* spp.)
    crosses
        composite, 64
        in landraces, 64–67
    landrace studies, *see* Barley in Ethiopia;
        Barley landraces in the Fertile
        Crescent
    regionality of resistance traits, 36
Barley in Ethiopia
    background, 77, 81
    categories and classifications
        hulled, 89–90
        hull-less, 90–91
        main groups, 87–89
        partially hulled, 91–92
    covariant characteristics, 85–87
    crop description, 79–80
    diversification factors, 92–94
    ex situ conservation, 78, 97, 101–102
    geographic distribution, 83–85
    history
        cultivation and use, 80–81
        exploration and studies, 81–83
    improvement programs
        modern breeding systems, 95–96
        natural/traditional breeding
            systems, 94–95
    significance of, 85
    in situ conservation
        changes in, 100–101
        description and background, 97–98
        farming partners project, 102–103
        landrace restoration effort, 103
        principles of integration with *ex situ*,
            101–102
        programs for, 99–100
        relevance to genetic diversity, 98–99